S0-BLI-460

Chemical Generation and Reception of Radio- and Microwaves

Chemical Generation and Reception of Radio- and Microwaves

Anatoly L. Buchachenko
Eugene L. Frankevich

Anatoly L. Buchachenko
Institute of Chemical Physics
Academy of Sciences
University of Moscow
117977 Moscow
Russia

Eugene L. Frankevich
Institute of Chemical Physics
Academy of Sciences
117977 Moscow
Russia

This book is printed on acid-free paper. ♾

Library of Congress Cataloging-in-Publication Data

Buchanchenko, A. L. (Anatoly Leonidovich)
Chemical generation and reception of radio and microwaves/
Anatoly L. Buchachenko, Eugene L. Frankevich.
p. cm.
Includes index.
ISBN 1-56081-630-9
1. Quantum chemistry. 2. Nuclear magnetism. 3. Chemical reactions. I. Frankevich, Eugene L. II. Title.
QD462.B83 1994
541.2'8--dc20 93-28335
CIP

Printed in the United States of America

ISBN 1-56081-630-9 VCH Publishers
ISBN 3-527-89630-9 VCH Verlagsgesellschaft

Printing History:
10 9 8 7 6 5 4 3 2 1

VCH Publishers, Inc.
220 East 23rd Street
New York, New York 10010

VCH Verlagsgesellschaft mbH
P.O. Box 10 11 61
69451 Weinheim
Germany

VCH Publishers (UK) Ltd.
8 Wellington Court
Cambridge CB1 1HZ
United Kingdom

To the memory of

GERHARD L. CLOSS,

the Great Man in Spin Chemistry

Preface

This book elucidates two new and remarkable facets of chemical reactions not too familiar to the chemical and physical scientific community.

At certain conditions a chemical reaction is a generator, an emitter of coherent electromagnetic waves, which it emits like a tiny molecular broadcasting station, similar to a chemical maser or laser. On the other hand, electromagnetic waves affect chemical, and some biochemical processes, modify the chemical reactivity of paramagnetic reagents, and, consequently, result in new phenomena—radio-induced magnetic isotope effects, quantum beats in chemical reactivity, new types of stimulated alignment of magnetic nuclei, and new principles of electron spin resonance detection.

This is the reason why it might be said that this book formulates *chemical radiophysics* as a new field of chemical physics. In chemical radiophysics both the generation and reception of electromagnetic waves occur due to the magnetic component of the electromagnetic field, whereas the key role in radiophysics generally belongs to the electric component.

The foundation of chemical radiophysics is *spin chemistry*, a new branch of chemistry that analyzes electron and nuclear spin angular momenta behavior in chemical processes and explores the chemical reaction magnetic scenario, written by magnetic interactions. It examines the spin-mediated magnetic effects and formulates spin-related mechanisms of the electromagnetic field effects in chemistry and biology.

Spin chemistry has been successfully developed for the past two decades by the efforts of many eminent scientists. J. Bargon, H. Fischer, R. Lawler, G. Closs, R. Kaptein, Yu. Molin, K. Salikhov, R. Sagdeev, N. Turro, P. Atkins, K.

McLauchlan, and H. Roth are the members of the international team that has played a key role in laying down the foundations of this science. The authors of this book have every reason to believe that they also have the honor of belonging to this team, so the book is written by people who are not detached onlookers in spin chemistry.

The intellectual potential of spin chemistry, and chemical radiophysics in particular, manifests itself and is promising in many aspects. First, it makes a new contribution in understanding the physics of chemical reactions and offers new insight into chemical dynamics and reaction mechanisms, so we believe the book will be useful to chemists and chemical physicists. Second, chemical radiophysics provides new radiospectroscopic methods and techniques of exquisite sensitivity to paramagnetic intermediates. Therefore, the book might be of interest to physicists and radiospectroscopists. Third, there is much speculation about biological electromagnetic field effects, so we intend to discuss them with appeals to spin-mediated mechanisms in order to distinguish real knowledge from irrelevant speculation; therefore, we hope that the book might be attractive to biologists and biophysicists.

Acknowledgments

We are gratefully indebted to VCH Publishers, particularly to Dr. Charles H. Doering, for friendly cooperation, to Tatyana A. Sapego for her help with the English translation, and to Alexei A. Buchachenko for his valuable assistance in manuscript preparation.

Anatoly L. Buchachenko
Eugene L. Frankevich
Moscow
June 1993

Contents

CHAPTER

1

Magnetic Scenario of Chemical Reaction

Magnet, magnetism, and magnetic field are part of modern civilization. Every discovery about the magnet, starting with that of the compass needle, has been a leap forward and accelerated the evolution of civilization. Electricity and electric engines, modern energy sources, communication, radio and TV, computers—these are the offspring of the magnet. Everyone realizes the place and great importance of the magnet in physics, geography, geology, technology, and even in archeology. What about chemistry?

1.1. The Energy Dogma in Chemistry

Any chemical event, its feasibility and result, is traditionally rationalized according to its energetic scenario: how much energy it requires, how to pump this energy to activate the reacting species, and how efficiently this energy is adopted in chemically reacting system.

Chemical reaction is indeed the *physical process* of rearranging electrons and atoms, resulting in the transformation of molecules. The great American physicist R. Feynman once remarked in despair that chemistry is the most complicated physics, the highest physics, unattainable for physicists to understand. The reason is that too many electron–electron and electron–nuclear time-dependent interactions are involved in a chemical event.

The chemical transformation of substances and materials is an energy-expensive process and, hence, all principles of chemical reaction control and stimulation come from the energy dogma (i.e., from the idea of overcoming an energy deficit either by searching for the ways to reduce energy expenses of reactions or by securing efficient routes for energy pumping the reacting molecules). This principal strategy has ensured the brilliant successes of chemistry in materials science, in modern chemical technology, and in chemical energetics.

Among the variety of methods of stimulating chemical reactions, based on the energy dogma, the action of magnetic fields has never been considered seriously for an obvious reason: the additional magnetic energy of reacting species, molecules, and their fragments—atoms, ions, and radicals—is negligible even in the strongest magnetic fields attainable. It is millions of times smaller than thermal energy and hundreds of millions of times smaller than that necessary for a chemical reaction. That is why the combination of the words *magnet* and *chemistry* is at least naive in terms of energy dogma.

However, in addition to energy there is another physical property that is of paramount importance for chemistry—the *angular momentum* (*spin*) of electrons and nuclei of reagents. All chemical reactions are spin selective (i.e., they are allowed only from the specific spin states and are prohibited from others).

Such a spin selectivity of chemical events implies that chemically identical reagents possess different, spin-dependent chemical reactivity. Only spin allowed reaction channels are open, the others, even if they are energetically permissible, are strictly closed for the reaction.

The only interactions that are able to disturb the spin of reagents and transform nonreactive, spin forbidden reagent state into the reactive, spin allowed ones, are the magnetic interactions. They contribute nothing to the total energy (and, therefore, may be completely ignored in terms of energy dogma), however, they control the spin behavior of reagents, they change their spin, and they switch the reaction from the spin forbidden channels to spin allowed ones (or vice versa). Ultimately, they modify chemical reactivity and create a new, *magnetic scenario* of chemical reaction.

Among the chemical species, the bearers of electron spin, the most famous and widespread are organic radicals, ion radicals, inorganic ions, many of atoms and excited molecules. The radical, as chemists know, is the part of the molecule formed by chemical bond scission in such a way that the pair of bonding electrons is shared equally between the molecule fragments, each of which acquires the unpaired electron. Radicals are formed by dissociation of molecules (for instance, hydrogen molecule can be split into two hydrogen atoms each possessing a single spin bearing unpaired electron; the ethane molecule can be split into two methyl radicals, $CH_3CH_3 \rightarrow \dot{C}H_3 + \dot{C}H_3$, etc.), by addition of atoms or other radicals to double chemical bond (for instance, hydrogen atom addition to ethylene yields ethyl radical, $CH_2CH_2 + H \rightarrow \dot{C}_2H_5$), and by many other ways. Attachment or detachment of single electron to a molecule produces the charged anion or cation radical, respectively; in addition the excitation of molecules into the triplet state generates chemical species that carry two unpaired electrons and two electron spins, respectively. Inorganic paramagnetic ions such as Fe^{3+}, Fe^{2+}, Cu^{2+} (but not Cu^{+}), and uranoyl ion UO_2^+ (but not uranyl ion UO_2^{2+}) are the typical and well-known chemical species—spin bearers. Many examples of their reactions and spin behavior will be given later.

1.2. Spin and Spin States

In contrast to macroscopic physical bodies (such as a wheel or a top), whose angular momentum can be of any magnitude, the momentum of microparticles takes only selected discrete values. The length of the electron spin angular momentum vector **S** is equal to

$$|\mathbf{S}| = \hbar[S(S+1)]^{1/2}$$

where S is the spin quantum number, $S = 1/2$, $\hbar = h/2\pi$, h is the Plank constant. The nucleus has the spin

$$|\mathbf{J}| = J = \hbar[I(I+1)]^{1/2}$$

where I is the nuclear spin quantum number. For the proton, neutron, nuclei ^{13}C, ^{15}N, ^{31}P $I = 1/2$, for D, ^{14}N $I = 1$, for nucleus ^{17}O $I = 5/2$, for nuclei ^{12}C, ^{16}O $I = 0$, and so on.

Usually, the term spin refers to the spin quantum number S rather than to the length of spin angular momentum vector. For example, when one says that the electron spin is equal to 1/2, it actually means that the spin quantum number $S = 1/2$, whereas $|\mathbf{S}| = (\sqrt{3/2})\hbar$. The same is true for nuclear spins.

The spin projection onto the selected quantization axis z also has discrete values, but is arbitrary for perpendicular directions x, y. For the electron

$$S_z = m\hbar$$

where $m = \pm 1/2$. Evidently, the spin S and its projection S_z are not equal (i.e., the spin is never directed along the quantization axis but makes an angle α with it such that $\cos\alpha = S_z/S$). It means that the spin precess about the direction of the axis z similar to the rotating top axis precessing around the vertical to form a rotation cone.

In multispin systems this may result in a remarkable consequence—the phase coherence in the precession [i.e., the collective in-phase motion, of individual spins that provides the *coherence* of chemical reactivity (quantum beats, see Chapter 11)].

While possessing a spin, electron and nuclei also have magnetic moment whose value and direction is unambiguously connected with those of spin. The magnetic moment of the electron is

$$\mu_e = -g(\beta/\hbar)\mathbf{S}$$

where β is an elementary "quantum" of magnetism, the Bohr magneton, g is the electron g-factor (i.e., the ratio of the magnetic moment of the electron to its angular momentum). For the free electron $g = 2$ (or, more precisely, taking into account the relativistic correction, $g = 2.0023$). In radicals the anisotropy of the electron distribution induces the angular momentum of orbital electron motion in addition to electron spin itself. Because of this extraorbital magnetism the g-factors of radicals are usually not equal to 2 and range from 1.9 to 2.1.

The projections of the electron magnetic moment μ_{ez} on the quantization axis similar to the spin projection have discrete values

$$\mu_{ez} = -g(\beta/\hbar)S_z = -\ g\beta m$$

where $m = \pm 1/2$. These projections correspond to both orientations of the electron spin (and the magnetic moment): along the direction of the quantization axis and opposite to it.

Similarly, the magnetic moment of the nucleus corresponds with its spin:

$$\mu_n = -g_n(\beta_n/\hbar)\mathbf{J}$$

where β_n is the Bohr nuclear magneton (i.e., a "quantum" of nuclear magnetism), while g_n is a nuclear g-factor, the value and sign of which are specific for each nucleus and depend on the inherent nuclear structure.

The projections of the nuclear magnetic moment also have discrete values:

$$\mu_{nz} = -g_n(\beta_n/\hbar)J_z = -g_n\beta_n m_I$$

where $m_I = 0, \pm 1, \ldots, \pm I$ defines the number of nuclear spin states. For a proton $I = 1/2$ and the only allowed values $m_I = \pm 1/2$ describe two possible orientations of magnetic moment—up and down relative to the quantization axis.

The same holds for the electron: the spin momentum may possess two projections $m = \pm 1/2$, which differ in the direction with respect to the z axis. The spin quantum number S and its projection m specify the pure spin state of the electron. The number of spin states is the spin multiplicity, which is expressed as $2S + 1$, where S is a spin quantum number. For electron $S = 1/2$ the spin multiplicity is equal to 2, that is, it is equal to the number of spin projections on the quantization axis.

Similarly, for the nucleus with spin quantum number I the nuclear spin multiplicity, that is the number of allowed nuclear spin projections, is $2I + 1$.

In diamagnetic molecules the electron spins are oriented opposite to each other and both total electron spin and its projections are equal to 0. The spin multiplicity of these molecules is equal to 1, so only one spin state, the singlet state S, is allowed. In radicals there is one electron with unpaired spin, therefore, the radical is allowed to be in two spin states, the components of doublet state D whose spin multiplicity is equal to 2.

An oxygen molecule and many electronically excited molecules have two unpaired electrons and total electron spin number $S = 1$. Their spin multiplicity is equal to $2S + 1 = 3$, that is, the total spin is oriented in three different ways with the projections $+1$, 0, and -1. The electron spin state of such molecules is said to be triplet (T). For a molecule with an electron spin of 3/2 the spin multiplicity is equal to 4 (i.e., its spin state is quartet, for a molecule with a spin of 2 the spin state is quintet, and so on).

The organization of electron spins determines the spin state. The total spin of two radicals (particularly, in the radical pair, one of the key intermediates of radical reactions) can be equal to 1 (the spins of partners are summed up) or to 0 (the spins are compensated). Accordingly, the radical pair can be in triplet T or singlet S spin states. The scheme of the spin arrangement in the radical pair is shown in Fig. 1.1. Three triplet spin states that differ by the total spin projections, equal to $+1$, 0, and -1, are designated as T_+, T_0, and T_-. In the singlet state S two spins of partners cancel each other as depicted in Fig. 1.1.

In the pair of radical and triplet molecule (for instance, in partnership of radical with oxygen, triplet carbene, or with other electronically excited triplet molecules) the total spin can be equal either to 3/2 or to 1/2. Accordingly, the pair can be in quartet, Q, or in doublet, D, states, with spin projections $\pm 3/2$, $\pm 1/2$ for Q and $\pm 1/2$ for D states, respectively. This scheme of the total spin construction and prediction of the spin states is easily generalized for any multispin molecular system.

It is clear that the different electron and nuclear spin states correspond to different spin and magnetic moment projections. Hence, these states acquire different magnetic energies in external magnetic field (i.e., the degeneracy on the projections disappears). The energy gap between the spin state components induced by magnetic field is the well-known Zeeman splitting.

It is worth remembering that the transitions between electron spin states (and, consequently, between the Zeeman sublevels) are responsible for electron spin resonance (ESR), while the transitions between the nuclear spin states form the basis for nuclear magnetic resonance (NMR). These transitions are induced by a microwave magnetic field at the frequency of electron or nuclear precession and are accompanied by spin projection changes. The magnetic field may be applied externally, as in ESR or NMR, or be produced by the lattice

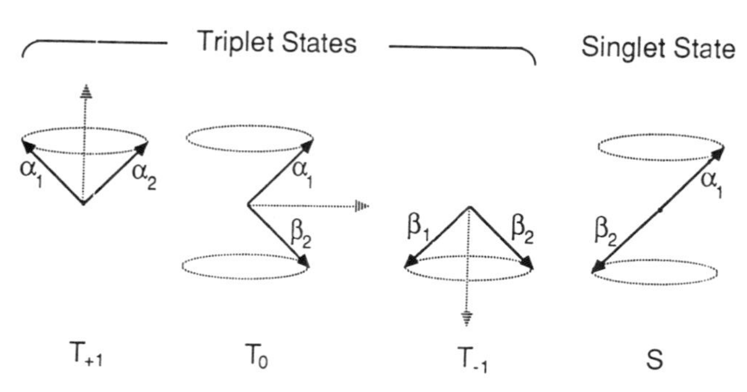

Figure 1.1 The scheme of the spin orientations for two unpaired electrons in triplet states T_+, T_0, T_-, and in singlet state S. The spin projections in these states are $+1$, 0, -1, and 0, respectively.

stochastic molecular motion, which produces the fluctuating, noisy magnetic fields of different frequencies and amplitudes. The spectral component of this noise at the frequency of electron or nuclear precession induces the spin–lattice or spin–spin relaxation (i.e., transitions between spin states).

1.3. Spin Selectivity of Chemical Reactions

The fundamental principle of chemical reactions is that both spins and their projections remain unchanged in elementary chemical events.

The conservation of the total spin results in an important consequence: *chemical reactions are spin selective, allowed only for such spin states of products whose total spin is identical to that of reagents and strictly forbidden if they require spin change.*

For example, the reactions

$$NH_3 + H^+ \rightarrow NH_4^+$$

$$CH_3Cl + HO^- \rightarrow CH_3OH + Cl^-$$

$$O(^1D) + H_2 \rightarrow H_2O$$

$$(CH_3)_2C{=}C(CH_3)_2 + {}^1O_2 \rightarrow (CH_3)_2\underset{|}{C}{-}\underset{|}{C}(CH_3)_2 \quad (\text{ring: } O{-}O \text{ bridging the two C atoms})$$

are spin allowed because the total spins of reagent molecules and that of products coincide, they are equal to zero for all examples including the last one, the transformation of 2,3-dimethylbutene into oxetane in the reaction with electronically excited singlet, zero spin oxygen molecule.

For the same reason the following reactions are also spin allowed:

$$R\dot{C}O \rightarrow \dot{R} + CO$$

$$CH_2{=}CH_2 + \dot{H} \rightarrow CH_3\dot{C}H_2$$

$$Cu^{2+}(H_2O)_6 + NH_3 \rightarrow Cu^{2+}(NH_3)_6$$

but now the total spins of reagents and products both equal 1/2.

From the two reactions

$$O(^3P) + H_2 \rightarrow HO + O$$

$$O(^3P) + H_2 \rightarrow H_2O$$

the first one is spin allowed, while the second is spin forbidden because the spin of reagents ($S = 1$) is not equal to that of the product molecule ($S = 0$).

The chemical coupling in the radical pair, either recombination or disproportionation, resulting in formation of diamagnetic molecules (e.g., of H_2 from two hydrogen atoms, of ethane from two methyl radicals) is spin allowed only from the singlet state of the pair. Triplet radical pairs do not react, until its spin state changes to a singlet state, from which the reaction starts generating the product molecules in the singlet state. The same is valid for the dissociation reaction: a zero-spin molecule is not allowed to generate a triplet radical pair; it can be split into the singlet pair only.

The reaction $H + O_2 \rightarrow HO_2$ (or the addition of any radical to oxygen or other triplet molecules) takes place in the doublet spin state of a reagent pair ($H\ O_2$) since the total spin of the pair in this state ($S = 1/2$) is identical to that of the reaction product, HO_2 radical. However, the reaction is forbidden in the quartet state of the pair since in this state its total spin ($S = 3/2$) is not identical to that of the HO_2 radical. Again, among two reaction channels only one is spin allowed.

The precursor for the reaction between two triplet molecules is a pair with nine conceivable spin states that differ in spin arrangement: quintets with total spin $S = 2$ and spin projections ± 2, ± 1, 0; triplets with $S = 1$ and projections ± 1, 0, and singlet $S = 0$. Chemical reactions differentiate these states and select only spin allowed ones. For example, the recombination reaction of two triplet carbenes Ph_2C into tetraphenylethylene

$$2\,Ph_2C \rightarrow Ph_2C{=}CPh_2$$

is allowed only from singlet (i.e., only one of nine possible spin states of the pair of carbenes is ready for reaction, the others are forbidden to react).

The addition of the ground state triplet oxygen molecule to the excited triplet anthracene, which yields the endoperoxide singlet molecule, also takes place from only one of nine spin states of the precursor pair.

The reaction of triplet carbene with oxygen

$$CH_2(S = 1) + O_2(S = 1) \rightarrow \dot{C}H_2O\dot{O}$$

can be directed into two channels. The first one results in the formation of the $\dot{C}H_2O\dot{O}$ biradical in the singlet state ($S = 0$) and, therefore, is allowed from only one of nine conceivable spin states of the prereactive ($CH_2\ \ O_2$) of the same spin multiplicity, while the second channel yields the same biradical in the triplet state ($S = 1$) and it is spin allowed from three of nine states of the pair. All other spin states are closed to reaction.

Biologically important chemical reactions are also spin selective and, therefore, are magnetosensitive. First, it concerns reactions of photosynthesis: many important details of the mechanism of these reactions were furnished as a result of investigations on magnetic effects including microwave magnetic effects (Chapters 8 and 11). At the level of cell biology and phenomenological enzyme chemistry there have been many observations of magnetic field influence on the efficiency of enzyme processes (the competition of mitochondria oxidation and breathing, the ratio between oxidation and oxidative phosphorylation, the activity of peroxidases and other enzymes, etc.) (see, for example [1–4]). The magnetic sensitivity of these processes is evidence of their spin selectivity.

A detailed understanding of magnetic effects in these processes has not yet been achieved, but it is very likely that the key role belongs to spin selectivity of the Fe ion reactions. For example, in the electron transfer reaction

$$Fe^{2+}(S = 2) + O_2(S = 1) \rightarrow Fe^{3+}(S = 5/2) + O_2^-(S = 1/2)$$

the products can be found in states with a total spin of 3 or 2, while the total spin of reagents can be equal to 3, 2, or 1. It means that direct reaction is allowed only from two spin states, the third state with spin $S = 1$ is closed to reaction.

In the process

$$Cu^+(S = 0) + O_2(S = 1) \rightarrow Cu^{2+}(S = 1/2) + O_2^-(S = 1/2)$$

direct reaction is spin allowed but the reverse is allowed only from one spin state with $S = 1$ and forbidden from the spin state with $S = 0$.

The reduction–oxidation of the A substrates with cytochrome Fe ions is also spin selective:

$$A(S = 0) + Fe^{2+}(S = 2) \rightarrow A^-(S = 1/2) + Fe^{3+}(S = 5/2)$$

Direct reaction products may possess total spin 3 and 2, hence, the direct reaction is not limited by the spin selection rules; however, the reverse reaction is allowed only from one spin state with $S = 2$ and, therefore, is supposed to be magnetosensitive.

There are a great variety of such types of reactions in enzyme processes, in biological systems of electron transport, and in phosphorylation. It is a wide area for investigation in which the ideas and principles of spin chemistry are expected to be promising and instructive. An inspiring example when these hopes became justified is photosynthesis (Chapter 8).

For all the above-mentioned cases of spin selective reactions one common feature is typical: the total spin of reagents differs from that of the products. Such situations often occur in reactions involving radicals, triplet molecules, and transition metal ions. They are quite usual for reactions D + D (recombination or disproportionation of two radicals in doublet states), D + T (addition of radicals to triplet species, electron transfer from an ion radical to a triplet molecule or triplet ion, etc.), and T + T (interaction of two triplet molecules, oxidation of triplet carbenes by molecular oxygen, etc.).

Spin selectivity is inherent not only for chemical reactions, but also for physical processes, such as the excited triplet quenching by radicals and triplet–triplet annihilation (Chapters 5–8).

Two general requirements should be satisfied for the total spin of the reagent pair (D + D, D + T, T + T, etc.) to be changed and became adequate to that of the reaction products: first, the presence of magnetic interactions in the pair that impact on the spin behavior and, second, the survival of the pair for a time long enough for the magnetic interactions to be able to change the spin state of the pair. By the order of magnitude this time is comparable to an inverse value of the magnetic energy responsible for the spin state interconversion.

Besides these two fundamental requirements that develop the magnetic scenario of chemical reaction, one should admit the key importance of nonmagnetic exchange interaction between unpaired electrons of pair partners. This interaction determines the energy separation between spin states and, consequently, the possibility of spin state interconversion, which can occur only if the spin-state mixing magnetic interaction is of comparable size. If the exchange interaction is much greater than the magnetic one the probability of transitions between the spin states of the pair is strongly reduced and the spin forbiddance becomes more strict. In other words, a strong exchange interaction "fixes" spins, and the stronger this fixing, the more difficult it is to violate the spin in a multispin system.

References

1. Barnothy, M., ed. *Biological Effects of Magnetic Fields.* Plenum, New York, 1964.
2. Figueras Roca, F. *Ann. Chim.* **1967**, *2*, 255.
3. Atkins, P. W., Lambert, T. P. The effect of a magnetic field on chemical reactions. *Ann. Rep. Prog. Chem.* **1975**, *72*A, 67.
4. Grissom, Ch. B. Magnetic field effects on enzymatic reactions. In *Magnetic Field and Spin Effects in Chemistry*, book of abstracts. Konstanz, Germany, 1992.

CHAPTER

2

Magnetic Interactions in Chemical Reactions

Thus, we came to the conclusion that the spin change in the reagent pair is adequate to variations in chemical reactivity and modifications of yields of reaction products. The only interaction able to produce the spin state interconversion is the magnetic, whose rank in the energy hierarchy is lowest (Fig. 2.1). Its energy is negligibly small, four to nine orders of magnitude smaller than that of chemical bonds; however, its power and uniqueness are in its ability to govern the spin.

Now, the questions are how spin conversion occurs, what is its mechanism, and what is the place of magnetic interactions in this mechanism. We will answer these questions using the magnetically induced spin transformation in a radical pair (RP).

2.1. Radical Pair Dynamics

Radical pair is a dynamic system from which the radicals can diffuse apart, traveling randomly in space and time, and then return and reencounter. During these travels there occurs pair spin conversion, which is governed by magnetic interactions in radicals and results in either triplet–singlet transformation of the pair (if its starting state was triplet) or singlet–triplet transformation (for pairs with a singlet starting state).

To transform the triplet RP into a molecule, the time coincidence of at least four events is necessary.

First, the radicals leaving RP at the moment $t = 0$ should return and

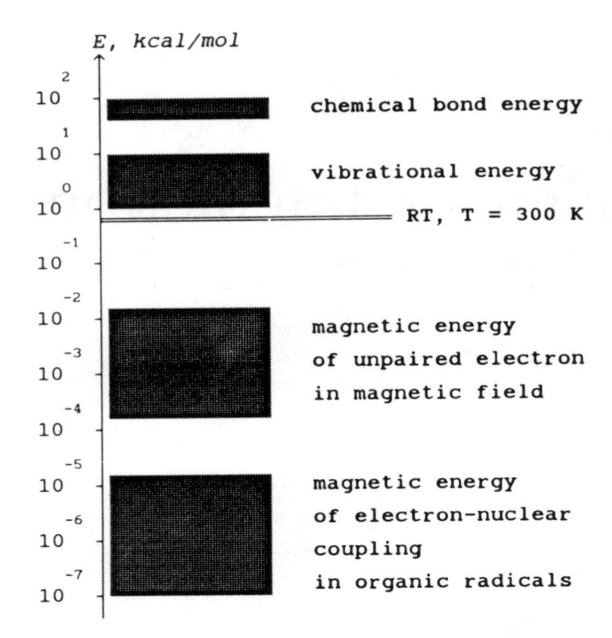

Figure 2.1 Energy scale for chemical and magnetic interactions.

reencounter at a certain moment *t*. This process of regeneration of contacts of RP partners is described by molecular or diffusion dynamics.

Second, at the same instant *t* the pair should urgently be in the singlet state and be ready to form a molecule. This process is controlled by spin dynamics, the dynamics of triplet–singlet conversion.

Third, during their diffusion motion the radicals may undergo chemical transformations (dissociation or scavenging) to yield a new RP instead of the initial one. For the primary molecule to be formed the pair is required to survive to time *t*. The probability of this event is determined by chemical dynamics.

Fourth, the pair being able to react (i.e., having survived to time *t* and being at this time in contact and in a singlet state) recombines into a molecule if a favorable orientation of the partners is achieved by rotational dynamics. Thus, the birth of a molecule in RP is the result of the collective efforts and coordinated choreography of all three dynamics—*spin*, *molecular*, and *chemical*.

The dynamic scenario of RP behavior is a common one, irrespective of whether the pair is in liquid and its molecular dynamics is described by free, unrestricted diffusion, it is localized in a closed microreactor (micelle or zeolite cavity) and its diffusion is space limited, the diffusion dynamics develops in the Coulomb potential (for ion radical pairs), or the radical pair is locked in molecular solids, crystals, or glasses (e.g., an electron-hole pair) and its molecular dynamics is described by migration, by hopping of the electrons, radical anions, or holes. The only difference between these situations results from the distinct characters of molecular dynamics.

The key act of the magnetic scenario is spin dynamics, led by magnetic interactions, among which Zeeman electron and electron-nuclear, hyperfine, interactions are principal, responding for magnetic effects in chemical reactions. The magnetic dipolar interaction of electrons and the spin-rotational coupling induce the electron spin relaxation, which often makes a considerable and sometimes a predominant contribution to spin dynamics; however, in these cases magnetic effects turn out to be suppressed. From this point of view, the dipolar and spin-rotational interactions are harmful, provoking competitive spin interconversion processes through the dipolar and spin-rotational relaxations. Although they also depend on the magnetic field, this dependence is weak and loosely controlled so there is little chance of influence either by magnetic field or by radiowaves.

Spin dynamics is modulated by molecular dynamics and is time limited by chemical dynamics. Molecular dynamics randomly varies the distance between the radical partners in pair and, consequently, modulates exchange interaction $J(r)$, which, in turn, influences the rate of triplet–singlet conversion and finally the spin dynamics of the pair. The chemical reactions of radicals (addition, dissociation, hydrogen atom abstraction, etc.) kill the pair stopping its spin dynamics and replace it by a new pair with its own spin dynamics.

2.2. Spin Dynamics

To imagine spin dynamics let us consider the simplest radical pair ($H\dot{R}_1\ \dot{R}_2$) in which one of the partners contains the only magnetic nucleus, proton. Zeeman electron energy of the first radical in magnetic field H is equal to $(1/2)g_1\beta H$ (g_1 is a g-factor of the radical $H\dot{R}_1$), whereas that of the second radical is $(1/2)g_2\beta H$, the hyperfine interaction (HFI) energy in $H\dot{R}_1$ radical is am, where a is HFI constant and m is the proton spin projection on the direction of external magnetic field, $m = \pm 1/2$.

The scheme of electron spin orientations in RP has been depicted in Fig. 1.1 and is reproduced spatially more clearly in Fig. 2.2, where the energy levels of S and T pair spin states are also shown. In high magnetic field the levels T_+, T_0, and T_- are separated by Zeeman energy $1/2(g_1 + g_2)\beta H$, while in the low field limit Zeeman energy is small and the levels T_+, T_0, and T_- are almost degenerated. The exchange energy J splits the levels S and T_0 and, generally, should be taken into account in the spin dynamics. Here, however, for the sake of simplicity, we confine ourselves to the free RP in nonviscous liquids, where the exchange energy may be ignored.

The probability of the RP coupling reaction (recombination or any other) is proportional to the triplet–singlet conversion rate, which, in turn, is

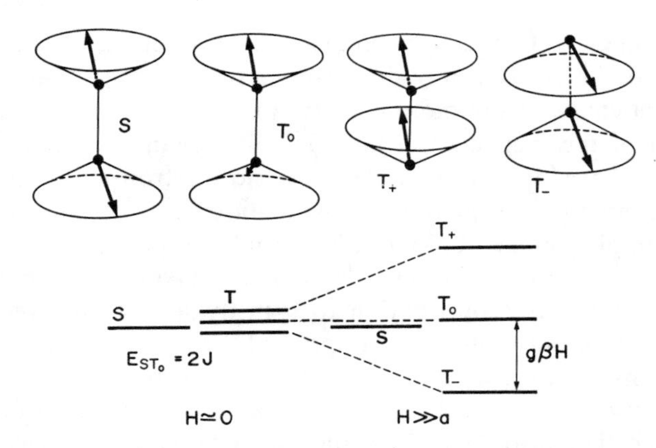

Figure 2.2 Energy diagram for the spin states of radical pair. In low magnetic field T states are almost degenerated, in high magnetic field $H \gg a$ Zeeman energy $g\beta H$ splits the T manifold into T_0, T_+, and T_- states. The S–T_0 gap E_{ST_0} is the exchange energy $2J$.

proportional to the squared matrix element $|\langle Tm|\mathscr{H}|Sm'\rangle|$ of the spin Hamiltonian

$$\mathscr{H} = g_1\beta H\mathbf{S}_1 + g_2\beta H\mathbf{S}_2 + a\mathbf{IS}_1$$

evaluated with the spin wave functions of S and T states. Here the spin Hamiltonian includes Zeeman and HFI terms only, $\mathbf{S}_1$ and $\mathbf{S}_2$ are the spin momenta of unpaired electrons, $\mathbf{I}$ is the proton spin, and m, m' are its projections.

Skipping over the details of calculations, which are simple and can be found in any textbook on quantum mechanics or radiospectroscopy [1–3], we write the results

$$\langle T_0 m|\mathscr{H}|Sm\rangle = (1/2)(\Delta g\beta H + am), \qquad m = \pm 1/2 \tag{2.1}$$

$$\langle T_+(-1/2)|\mathscr{H}|S(+1/2)\rangle = -(1/8)^{1/2}a \tag{2.2}$$

$$\langle T_-(+1/2)|\mathscr{H}|S(-1/2)\rangle = (1/8)^{1/2}a \tag{2.3}$$

where $\Delta g = g_1 - g_2$. As follows from Eq. (2.1), T_0–S transitions are induced by both Zeeman and HFI interactions. Alternatively, the $T_\pm$–S transitions are allowed only in the presence of HFI interactions and are inevitably accompanied by changes in the nuclear spin projection:

$$T_+(-1/2) \rightarrow S(+1/2), \qquad T_-(+1/2) \rightarrow S(-1/2)$$

or

$$T_+\beta_n \rightarrow S\alpha_n, \qquad T_-\alpha_n \rightarrow S\beta_n$$

where α_n, β_n designate the nuclear spin projections $m = \pm 1/2$. In other words, for the $T_\pm$–S transitions to take place the spins of electron and nucleus should be necessarily bound by the HFI interaction into a common spin system, so that the change of the former compensates that of the latter.

Equations (2.1)–(2.3) are obtained in terms of the elementary quantum mechanics, however, their classical analogs can be derived from the following simple and evident physical considerations. The precession frequency of the first electron (on the radical $H\dot{R}_1$) is equal to $(1/2)(g_1\beta H + 1/2a)$ if $m = +1/2$, and to $(1/2)(g_1\beta H - 1/2a)$ for $m = -1/2$; the precession frequency of the second electron (on the radical $\dot{R}_2$) is $(1/2)\beta H$. The difference in the frequencies of these precessions is therefore

$$\Delta\omega_m = (1/2)(\Delta g\beta H + am)$$

The T_0–S transition is equivalent to dephasing of two spins through the angle π (Fig. 2.3 a) and occurs at the condition $\Delta\omega_m t_m = \pi$ where t_m is a dephasing time, the intrinsic time of T_0–S transformation of pair

$$t_m \approx (\Delta g\beta H + am)^{-1}$$

It depends on the magnetic field, nuclear spin orientation, and HFI energy in accordance with the predictions of Eq. (2.1). At typical values of g-factors of organic radicals and HFI constants this time is 10^{-10}–10^{-7} s.

The interchange by angular momenta between the electron and nuclear spin subsystems can be also represented classically in terms of the spin vector model (Fig. 2.3 b). Let the spins S_1 and S_2 of both unpaired electrons of the pair be oriented in the same way (as in T_+ or T_- states) and one of the radicals has a nucleus with the spin I, which is oriented up. In zero magnetic field the electron S_1 and the nuclear I spins in the radical are summed up and precess

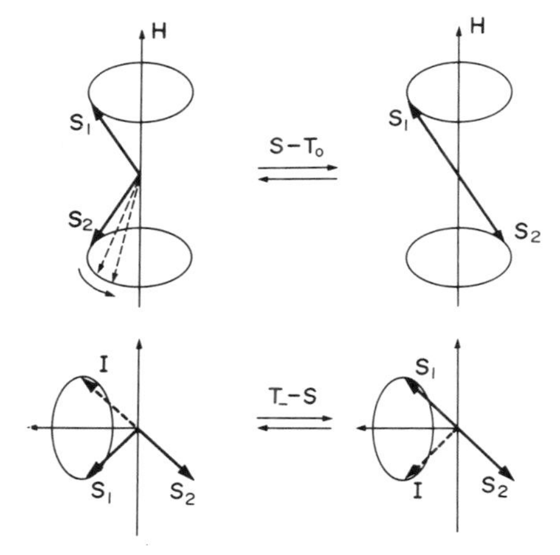

Figure 2.3 Schematic representation of the S–T_0 spin conversion induced by the dephasing of two electron spins $\mathbf{S}_1$ and $\mathbf{S}_2$ (top) and the scheme of the angular momenta interchange between electron $\mathbf{S}_1$ and nuclear $\mathbf{I}$ spin subsystems, T_-–S transition (bottom).

around the direction of the total spin vector. As a result of the dephasing the electron and nuclear spins exchange their orientations: now electron spins S_1 and S_2 compensate each other (this corresponds to the singlet state of the pair), while the nuclear spin is oriented down. This electron–nuclear flip-flop process corresponds to the $T_-\alpha_n \rightarrow S\beta_n$ transition. The $T_+\beta_n \rightarrow S\alpha_n$ transition can be constructed as an exact mirror reflection of the $T_-\alpha_n \rightarrow S\beta_n$ spin conversion.

There is no problem generalizing Eqs. (2.1)–(2.3) to the case of a radical pair with any number of different nuclei. The spin Hamiltonian in this case should be extended as

$$\mathscr{H} = g_1\beta H\mathbf{S}_1 + g_2\beta H\mathbf{S}_2 + \sum_i a_i\mathbf{S}_1\mathbf{I}_i + \sum_j a_j\mathbf{S}_2\mathbf{I}_j \tag{2.4}$$

where $\mathbf{I}_i$ is the spin of the ith sort of nuclei in the first radical, $\mathbf{I}_j$ is that of the jth sort of nuclei in the partner, and a_i and a_j are the HFI constants. Hence

$$\langle \mathrm{T}_0 m|\mathscr{H}|\mathrm{S}m\rangle = (1/2)\left(\Delta g\beta H + \sum_i a_i m_i^{\mathrm{a}} - \sum_j a_j m_j^{\mathrm{b}}\right) \tag{2.5}$$

where m_i^{a} and m_j^{b} are the projections of the ith sort of nuclear spins in the first radical and that of the jth sort of nuclear spins in the partner, respectively. Clearly, the rate of T_0–S transition is proportional to $\Delta g\beta H$ and depends on the nuclear spin orientations.

The rate of the $\mathrm{T}_\pm$–S transitions is proportional to the matrix elements

$$\langle T_+, m_i - 1|\mathscr{H}|\mathrm{S}, m_i\rangle = -(1/8)^{1/2}a_i[I_i(I_i + 1) - m_i(m_i - 1)] \tag{2.6}$$

$$\langle \mathrm{T}_-, m_i + 1|\mathscr{H}|\mathrm{S}, m_i\rangle = (1/8)^{1/2}a_i[I_i(I_i + 1) - m_i(m_i + 1)] \tag{2.7}$$

These transitions are accompanied by nuclear spin reorientation: at T_+–S conversion the nuclear spin increases ($m_i - 1 \rightarrow m_i$), while at T_-–S conversion it decreases ($m_i + 1 \rightarrow m_i$).

It follows from Eqs. (2.1)–(2.3) and (2.5)–(2.7) that the probability of a triplet–singlet conversion and, consequently, the probability of chemical reaction in RP depends on

1. The magnetic field strength H through the difference in electron Zeeman energies $\Delta g\beta H$;
2. The nuclear magnetic moments through the HFI energy a, as far as

   $$a = (8\pi/3)\mu_{\mathrm{e}}\mu_{\mathrm{n}}|\psi(0)|^2$$

   where μ_{e} and μ_{n} are magnetic moments of the electron and nucleus and $|\psi(0)|^2$ is the electron spin density at the nucleus;
3. The nuclear spin orientations m^{a} and m^{b}.

The RP reaction probability might be represented in the closed form as

$$P = F(H, a, m)$$

This compact formula contains the main magnetic effects in chemical reactions:

magnetic field effect, $P(H)$, magnetic isotope effect, $P(a)$, and chemically induced dynamic nuclear polarization, $P(m)$.

The general theory of RP reaction dynamics that integrates the spin, molecular, and chemical dynamics is presented in the next section.

2.3. Theory of the Extended Spin Dynamics

The key problem of the theory is the calculation on the probability of chemical coupling in the radical pair.

There are two levels of approximations, which we will classify as nonempirical and semiempirical. In terms of the nonempirical theory the molecular and spin dynamics are treated commonly as combined dynamics; the semiempirical theory ignores their correlation and treats them independently.

2.3.1. Semiempirical Theory

Exchange interaction, $J(r)$, is a short-range potential that rapidly decays at the distances of order of molecular diameter. In other words, it affects only very short trajectories of diffusion motion of radicals in pair. The time that the pair spends along these trajectories is short, significantly shorter than that of triplet–singlet conversion. Therefore, as a rule, the short diffusion trajectories are inefficient for spin conversion. The overwhelming contribution into the triplet–singlet conversion is due to longer and continuous trajectories, but for these exchange interactions between far distant radicals in pair can be neglected. As a consequence, all three dynamics—spin, molecular, and chemical—can be treated independently.

The probability of molecule generation from RP in nuclear spin states *ab* is determined by the integral

$$P_{\mathrm{ab}} = \int_0^{\infty} \varepsilon |C_{\mathrm{S,ab}}(t)|^2 f(t) \exp(-kt)\, dt \tag{2.8}$$

The factor $|C_{\mathrm{S,ab}}(t)|^2$ describes spin evolution of the pair, being the probability of finding the pair in singlet state at time t. It can be evaluated as follows. The spin wave function Φ of RP may be represented as the product of nuclear χ_{ab} and electron spin functions; the latter is a linear combination of singlet $|\mathrm{S}\rangle$ and triplet $|\mathrm{T}\rangle$ spin functions

$$\Phi = [C_{\mathrm{S,ab}}(t)|\mathrm{S}\rangle + C_{\mathrm{T,ab}}(t)|\mathrm{T}\rangle]\chi_{\mathrm{ab}} \tag{2.9}$$

In high magnetic fields it is sufficient to take into account only one triplet substate, namely $\mathrm{T_0}$. Basic electron wave functions of pair are

$$|\mathrm{S}\rangle = (1/\sqrt{2})(\alpha_1\beta_2 - \beta_1\alpha_2), \qquad |\mathrm{T_0}\rangle = (1/\sqrt{2})(\alpha_1\beta_2 + \beta_1\alpha_2) \tag{2.10}$$

where $\alpha_{1,2}$ and $\beta_{1,2}$ are the spin functions of unpaired electrons located on radicals a and b. Spin evolution of the pair is obeyed by the time-dependent Schroedinger equation,

$$i\frac{\partial \Phi}{\partial t} = \mathcal{H}\Phi$$

where the Hamiltonian $\mathcal{H}$ includes both Zeeman and hyperfine interactions in the pair

$$\mathcal{H} = (g_a \mathbf{S}_a + g_b \mathbf{S}_b)\beta H + \sum_i a_i \mathbf{S}_a \mathbf{I}_i + \sum_j a_j \mathbf{S}_b \mathbf{I}_j \tag{2.11}$$

Solving the Schrodinger equation one can find the coefficients

$$C_{S,ab}(t) = C_S(0)\cos(Q_{ab}t) - iC_{T_o}(0)\sin(Q_{ab}t)$$

$$C_{T_o,ab}(t) = -iC_S(0)\sin(Q_{ab}t) + C_{T_o}(0)\cos(Q_{ab}t)$$

whose squares determine the probabilities for the pair to be in the singlet or triplet state. The squared coefficients $C_S(0)$ and $C_{T_o}(0)$ are the initial populations of the S and T_0 states and Q_{ab} are the nondiagonal $\mathcal{H}$ matrix elements:

$$Q_{ab} = \langle T_0 m|\mathcal{H}|Sm\rangle_{ab} = \frac{1}{2}\left((g_a - g_b)\beta H - \sum_i a_i m_i^a - \sum_j a_j m_j^b\right) \tag{2.12}$$

Since $\frac{1}{2}(g_a \beta H + \Sigma_i a_i m_i^a)$ and $\frac{1}{2}(g_b \beta H + \Sigma_j a_j m_j^b)$ are the frequencies of the unpaired electron spin precession in radicals a and b, then their difference $2Q_{ab}$ is the rate of dephasing, which is related to triplet–singlet T_0–S conversion.

If the starting state of the pair is singlet, i.e.

$$|C_S(0)|^2 = 1, \qquad |C_{T_o}(0)|^2 = 0$$

then

$$|C^S_{S,ab}(t)|^2 = \cos^2(Q_{ab}t) \tag{2.13}$$

if it is triplet,

$$|C_S(0)|^2 = 0, \qquad |C_{T_o}(0)|^2 = 1/3$$

then

$$|C^{T_o}_{S,ab}(t)|^2 = \tfrac{1}{3}\sin^2(Q_{ab}t) \tag{2.14}$$

where we have introduced the superscript indicating the initial state of RP.

Evidently, the spin state of RP oscillates between singlet and triplet with the characteristic time $t = \pi/Q_{ab}$, which ranges from 10^{-7} to 10^{-10} s, depending on the value of Q_{ab}. These oscillations, or quantum beats, were experimentally observed (Chapter 11).

Simultaneously, the spin dynamics is modulated by molecular dynamics described quantitatively by the function $f(t)$, which measures the probability that the radicals having escaped from the pair at $t = 0$ return and reencounter for the first time at instant t. In other words, $f(t)$ is the probability of the first diffusion contact; the starting contact of radicals in pair at $t = 0$ is identified as the zero one.

The popular and frequently applied form of the $f(t)$ function is that suggested by Noyes [4]:

$$f(t) = mt^{-3/2}\exp(-\pi m^2/p^2 t), \tag{2.15}$$

where $p = \int_0^\infty f(t)\,dt$ is the total probability that the radicals will reencounter at least once during the RP lifetime; then, $(1 - p)$ is the probability that radicals will never return and meet again. The approximate expressions for p and the parameter m were given by Noyes:

$$p \simeq 1 - (1/2 - 3\rho/2\sigma)^{-1} \tag{2.16}$$

$$m \simeq 1.036(1 - p)^2(\rho/\sigma)^2 v^{1/2} \tag{2.17}$$

where ρ is the diameter of contact pair, σ is the distance of the root-mean-square diffusional displacement of radicals ($\sigma^2 = 6D/v$), v is the frequency of the diffusion jumps, and D is diffusion coefficient. Assuming for the semiquantitative estimations $\rho = \sigma$ and $v \simeq 10^{-12}\,\mathrm{s}^{-1}$ we find $p \approx 0.5$ and $m \approx 10^{-6}$–$10^{-7}\,\mathrm{s}^{-1/2}$. For times $t \gg m^2/p^2 \approx 10^{-12}\,\mathrm{s}$, $f(t)$ can be approximated by $mt^{-3/2}$.

The function $f(t)$ describes the time distribution of the diffusion trajectories of radicals. The reencounter probability is high for short trajectories, but decreases as far as trajectories become longer. However, in the long trajectories, spin evolution and efficiency of triplet–singlet conversion are complete. The most favorable situation occurs when the time of the first reencounter coincides with the time of triplet–singlet conversion.

The term $\exp(-kt)$ in Eq. (2.8) accounts for the chemical dynamics of the pair and measures the survival probability of the pair at time t; k is the rate constant of chemical transformation of the RP. Finally, ε is the probability that the radicals being in contact singlet pair react (i.e., the recombination cross section).

Equation (2.8) determines the probability P_{ab} of the molecule formation from RP in nuclear spin state ab at the first diffusion contact [5]; for triplet pair as a precursor

$$P^{\mathrm{T}}_{\mathrm{ab}} = (1/6)(y - q_{\mathrm{ab}}), \tag{2.18}$$

while for singlet pair

$$P^{\mathrm{S}}_{\mathrm{ab}} = (1/2)(y + q_{\mathrm{ab}}), \tag{2.19}$$

where

$$y = \exp[-(2m/p)(\pi k)^{1/2}] \tag{2.20}$$

$$q_{\mathrm{ab}} = \exp(-\Phi_{\mathrm{ab}})\cos(\Theta_{\mathrm{ab}}) \tag{2.21}$$

$$\Phi_{\mathrm{ab}} = (m/p)(2\pi)^{1/2}[(k^2 + 4Q^2_{\mathrm{ab}})^{1/2} + k]^{1/2} \tag{2.22}$$

$$\Theta_{\mathrm{ab}} = (m/p)(2\pi)^{1/2}[(k^2 + 4Q^2_{\mathrm{ab}})^{1/2} - k]^{1/2} \tag{2.23}$$

As a rule the rate constant k is considerably less than the frequency of diffusion jumps ($k \ll m^{-2}$) and therefore both $\Phi_{ab} \ll 1$ and $\Theta_{ab} \ll 1$. Then, Eqs. (2.20) and (2.21) can be reduced to

$$y = 1 - (2m/p)(\pi k)^{1/2} \tag{2.24}$$

$$q_{ab} = 1 - (m/p)(2\pi)^{1/2}[(k^2 + 4Q_{ab}^2)^{1/2} + k]^{1/2} \tag{2.25}$$

The total recombination probabilities of pairs P^S and P^T can be found by summation of P_{ab} over all nuclear spin states *ab*.

Equations (2.18) and (2.19) determine the probabilities of molecule formation at the first reencounter. For the triplet pair the contribution of the starting, zero contact vanishes because this pair can only dissociate, while the singlet pairs, even in zero contact, produce the molecules with probability ε and dissociate with probability $(1 - \varepsilon)$. Therefore, the probability of molecule formation from the singlet pair in both zero and first diffusion contacts together is equal to

$$P(S) = \varepsilon + (1 - \varepsilon) \sum_{ab} P_{ab}^S \tag{2.26}$$

However both zero and first contacts could fail to produce the molecule, hence, multiple reencounters (i.e. all subsequent contacts) must be taken into account.

First let us consider the singlet pair and assume that it does not undergo any spin evolution. The total probability of molecule formation in all successive contacts is the sum

$$P = \varepsilon + (1 - \varepsilon)p\varepsilon + (1 - \varepsilon)^2 p\varepsilon + \cdots = \frac{\varepsilon}{1 - p + \varepsilon p} \tag{2.27}$$

For the singlet pair in nuclear spin state *ab* with spin evolution, the probability for all contacts is equal to

$$P_{ab}(S) = \varepsilon + (1 - \varepsilon)P_{ab}^S + (1 - \varepsilon)(p - P_{ab}^S)P_{ab}^S + \cdots$$

$$= \varepsilon + (1 - \varepsilon)P_{ab}^S/(1 - p + P_{ab}^S) \tag{2.28}$$

and the total probability is

$$P(S) = \varepsilon + (1 - \varepsilon) \sum_{ab} P_{ab}^S/(1 - p + P_{ab}^S) \tag{2.29}$$

where the first term is the contribution of zero contact and the second one is that of all others.

For triplet pairs the total recombination probability in nuclear spin state *ab* due to multiple reencounters can be expressed by the sum

$$P_{ab}(T) = P_{ab}^T + (p - P_{ab}^T)\, P_{ab}^T + \cdots = P_{ab}^T/(1 - p + P_{ab}^T) \tag{2.30}$$

In Eqs. (2.29) and (2.30) P_{ab}^S and P_{ab}^T are determined by Eqs. (2.18) and (2.19).

In terms of semiempirical theory some parameters that are necessary to calculate the recombination probability either are known (Q_{ab}) or can be obtained experimentally (k); other parameters (e.g., m, p) can be estimated approximately. The only adjustable parameter of the theory that depends on the duration of contact, frequency of radical reorientation, and steric hindrance of the radical reaction center is ε.

Molecular dynamics of pairs in liquids has been discussed in detail by Razi Naqvi et al. [6], who presented the exact equation for function $f(t)$ different from that by Noyes. Belyakov and Buchachenko [7] have used their function to evaluate the RP recombination probability.

For two-dimensional molecular dynamics (diffusion in thin films, Langmuir–Blodgett molecular layers, etc.) the $f(t)$ function has been derived and discussed by Deutch [8]. For the molecular dynamics of ion radical pairs (diffusion in Coulomb potential) various functions $f(t)$ have been proposed (see, for instance, [9]). Molecular dynamics in a microreactor (confined diffusion in a sphere with reflecting walls) has been treated independently by Sterna et al. [10] and Tarasov and Buchachenko [11]. However, detailed analysis of the $f(t)$ function goes beyond the scope of this book.

2.3.2. Nonempirical Theory

The most rigorous theory of the RP recombination was formulated by Freed and Pedersen [12, 13]. It is based on the numerical solution of the stochastic Liouville equation (SLE), which describes both molecular and spin dynamics of a pair simultaneously, taking into account the exchange interaction and spin selective chemical reaction. The exchange potential J is assumed to be an exponential function of interradical distance r similar to that for a hydrogen-like molecule:

$$J(r) = J_0 \exp[-\lambda(r - d)] \tag{2.31}$$

where λ is a characteristic parameter and d is the distance of closest approach of radicals in the pair.

The most important numerical quantity is P, the probability of molecule formation during the pair lifetime. As was stated in the previous subsection, it depends on the starting spin state of the pair and can be expressed in terms of fundamental parameters of molecular and spin dynamics.

According to this theory the probability of molecule production from triplet RP is

$$P_{ab}^{T} = \frac{1}{3} \frac{\Lambda F}{1 + (1 - \Lambda)F} \tag{2.32}$$

where Λ is the probability of molecule production from the singlet pair, calculated without taking into account the singlet–triplet conversion and F is the probability of the singlet–triplet conversion. So, as a first approximation, the product ΛF is the probability of triplet–singlet transformation of the RP multiplied by the reaction probability of the singlet pair. The term $(1 - \Lambda)F$ in Eq. (2.32) takes into account the unfavorable chance that the pair, being transformed from triplet to singlet and being ready to react, avoids the reaction since $\Lambda < 1$.

The Freed–Pedersen theory was developed for high magnetic fields, and hence it refers only to the conversion between S and T_0 states. This is the reason why Eq. (2.35) includes a normalization factor of 1/3.

The probability of molecule generation from the singlet pair is

$$P_{ab}^{S} = \Lambda - \frac{\Lambda(1 - \Lambda)F}{1 + (1 - \Lambda)F} \tag{2.33}$$

where the first term is the probability of molecule formation from the singlet pair provided that there is no singlet–triplet evolution, and the second term is the probability of singlet pair leakage due to singlet–triplet conversion.

The quantity Λ can be estimated semiempirically or taken as an empirical parameter; the value F in the Freed–Pedersen theory is calculated as a function of the parameter $(Q_{ab}d^2/D)$, where Q_{ab} is determined by Eq. (2.12).

In the frame of nonempirical theory the effect of Coulomb potential on molecular and spin dynamics has been considered and an important self-consistent model was developed to account for the influence of the exchange spin-dependent interaction on both spin and molecular dynamics in the field of exchange potential [12, 13]. It has been shown that the Coulomb potential affects recombination probability more strongly than the self-consistent exchange potential does. The effect of electron relaxation in radicals was also included.

At last, it is noteworthy that several simplified versions of the nonempirical theory have been developed. Lawler et al. [14] and Habercorn [15] have also explored the SLE approach, but they approximated the exchange interaction by δ-function on the exchange sphere; Salikhov et al. [16, 17] have used a similar approach, but they treated the contribution of exchange interaction into singlet–triplet conversion as a perturbation.

To summarize, the nonempirical theory is important for understanding the physics of the spin effects and for prediction of their dependences on the physical factors, however, for semiquantitative estimations the semiempirical theory works quite satisfactorily.

Nevertheless, there are three instances in which the exchange interaction cannot be surely neglected and, therefore, the semiempirical theory should be used carefully. These are

1. Very viscous solutions, where a highly important contribution to singlet–triplet conversion comes mainly from the short diffusion trajectories in which radicals spend a long time because of the slow diffusion;
2. The cases of very high hyperfine coupling constants, which make singlet–triplet conversion extremely fast (of order 10^{-11}–10^{-12} s) within short diffusion trajectories where exchange interaction is fairly strong; similar behavior is inherent to pairs of radicals with short relaxation times;

3. The molecular systems with spatially restricted diffusion (biradicals with flexible bridges between radical centers, RP in micelles or in zeolite cavities). The theory for these systems has been put forward by Sterna et al. [10] and Tarasov and Buchachenko [11].

2.3.3. Spin Dynamics in Low Magnetic Fields

In low and zero magnetic fields, besides $S–T_0$ spin conversion, two other channels, $S–T_+$ and $S–T_-$, become important as well. The calculation of the RP recombination probabilities is not simple, since all three channels are not independent and, strictly, triplet–singlet conversion via these channels cannot be treated additively. However, as a first approximation, their interference may be neglected, so the above equations of nonempirical or semiempirical theories remain applicable to determine the low-field recombination probabilities. The only distinction between the channels lies in the values of Q_{ab}: for $S–T_0$ conversion it is given by Eq. (2.12), while for $S–T_{\pm}$ conversions the respective values are

$$Q_{ab}^{+} = -(1/8)^{1/2} a_i [I_i(I_i + 1) - m_i(m_i - 1)] \tag{2.34}$$

$$Q_{ab}^{-} = (1/8)^{1/2} a_i [I_i(I_i + 1) - m_i(m_i + 1)] \tag{2.35}$$

Again, it follows from these equations that $S–T_{\pm}$ transitions are allowed only for the nonzero HFI, and are accompanied by reorientation of nuclear spin.

The additivity assumption is rather poor and can be used only for semiquantitative speculations. For a particular case of RP with one nuclear spin I the problem of spin evolution in low and zero magnetic fields has been solved rigorously [3]. In zero field the evolutions of a singlet-state population for pairs prepared initially in triplet and singlet states are given by the functions

$$|C_S^T(t)|^2 = \frac{4I(I+1)}{(2I+1)^2} \sin^2[(2I+1)(a/2)t] \tag{2.36}$$

$$|C_S^S(t)|^2 = 1 - \frac{4I(I+1)}{(2I+1)^2} \sin^2[(2I+1)(a/2)t] \tag{2.37}$$

respectively. Substituting these populations in Eq. (2.8) one can find the recombination probabilities in zero magnetic field.

Nonempirical calculation on the probability of molecule production in low fields from RP with single magnetic nucleus $I = 1/2$ was given by Zientara and Freed [18], who have shown that these probabilities depend on HFI and Zeeman energies, but are only slightly affected by the exchange interaction. For multinuclear radical pairs a semiempirical statistical model to evaluate the low-field probabilities has been proposed [19]. A more detailed survey of theories and their various approximations was made recently by Steiner and Ulrich [20].

References

1. Carrington, A., McLachlan, K. *Introduction to Magnetic Resonance.* Harper & Row, New York, 1967.
2. Atkins, P. W. *Molecular Quantum Mechanics.* Clarendon Press, Oxford, 1970.
3. Salikhov, K. M., Molin, Yu. N., Sagdeev, R. Z., Buchachenko, A. L. *Spin Polarization and Magnetic Effects in Radical Reactions.* Elsevier, Amsterdam, 1984.
4. Noyes, R. M. *J. Chem. Phys.* **1954**, *22*, 1349.
5. den Hollander, J. A. *Chem. Phys.* **1975**, *10*, 167.
6. Razi Naqvi, K., Mork, K., Waldenstrom, S. *J. Phys. Chem.* **1980**, *84*, 1315.
7. Belyakov, V. A., Buchachenko, A. L. *Russ. J. Chem. Phys.* **1983**, *1385*, 1510.
8. Deutch, J. *J. Chem. Phys.* **1972**, *56*, 6076.
9. Mozumder, A. *J. Chem. Phys.* **1968**, *48*, 1659.
10. Sterna, L., Ronis, D., Wolfe, S., Pines, A. *J. Chem. Phys.* **1980**, *73*, 5493.
11. Tarasov, V. F., Buchachenko, A. L. *Russ. J. Phys. Chem.* **1981**, *55*, 1921.
12. Freed, J. H., Pedersen, J. B. *Adv. Magn. Res.* **1976**, *8*, 1.
13. Pedersen, J. B., Freed, J. H. *J. Chem. Phys.* **1974**, *61*, 1517.
14. Evans, G. T., Flemming, P. D. III, Lawler, R. G. *J. Chem. Phys.* **1973**, *58*, 2071.
15. Habercorn, R. *Chem. Phys.* **1977**, *19,* 165.
16. Salikhov, K. M., Sarvarov, F. S., Sagdeev, R. Z., Molin, Yu. N. *Kinet. Catal. (Russ.)*, **1975**, *16*, 279.
17. Sarvarov, F. S., Salikhov, K. M. *React. Kinet. Catal. Lett.* **1976**, *4*, 33.
18. Zientara, G. P., Freed, J. H. *J. Chem. Phys.* **1979**, *70*, 1359.
19. Schulten, K., Wolynes, P. *J. Chem. Phys.* **1978**, *68*, 3292.
20. Steiner, U. E., Ulrich, T. *Chem. Rev.* **1989**, *89*, 51.

CHAPTER 3

Magnetic Effects in Chemical Reactions

Spin selectivity of chemical reactions and spin dynamics are the source of two generations of magnetic effects: the *native* effects induced by the chemical reaction itself and *magnetic* effects *stimulated by electromagnetic waves*. The relation between these two generations is schematically shown in Fig. 3.1.

Three main effects originate from the dependence of the reaction probability in radical pair on three parameters: magnetic field, hyperfine interaction, and nuclear spin orientation. The action of the electromagnetic field (EMF) results in new effects and these are the subject of this book. Here we consider briefly the native magnetic effects to gain a clear understanding of the magnetic effects of the second generation.

3.1. Magnetic Field Effect

The empirical search for evidence of magnetic field effects (MFEs) on chemical processes has a long history. This search has usually been initiated by people who were not perfectly aware of, or even ignore, the important fact that magnetic energy is negligible in comparison with chemical energy. The desired effects were often "discovered," but subsequent, more rigorous and professional experimental investigations usually performed by serious scientists refuted them. This peculiar duel has continued for almost a century, but only after the discovery of spin phenomena has the MFE problem been placed on a firm scientific basis.

The old story was summarized in a series of comprehensive reviews [1–3].

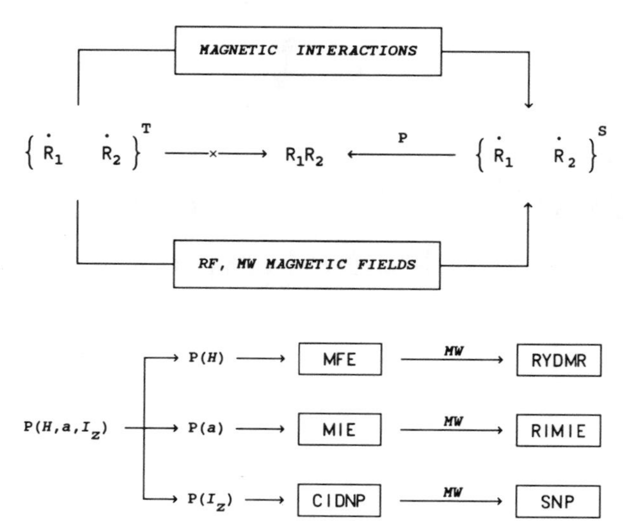

Figure 3.1 Two generations of magnetic effects. The first is induced by magnetic interactions, which transform the nonreactive triplet radical pair $|\dot{R}_1\ \dot{R}_2|^T$ into the reactive singlet pair $|\dot{R}_1\ \dot{R}_2|^S$. The probability of interradical reaction P is a function of magnetic field H, hyperfine interaction a, and nuclear spin orientation I_z, resulting in three first generation magnetic effects: magnetic field effect, MFE; magnetic isotope effect, MIE; and chemically induced dynamic nuclear polarization, CIDNP. Magnetic effects of the second generation are induced by radio-frequency or microwave magnetic fields and include reaction yield detected magnetic resonance, RYDMR; radio-induced magnetic isotope effect, RIMIE; and stimulated nuclear polarization, SNP.

A new spin history has been started by Frankevich, Geacintov, Avakian, Pope, and Merrifield et al. in 1965–1970 [4–10]. Spin selectivity of the molecular processes in triplet–triplet and triplet–doublet pairs has been considered by these authors and the influence of magnetic field on the spin statistics of these pairs was regarded as a source of MFEs in fluorescence quenching, in prompt and delayed fluorescence behavior, in fission and fusion processes of triplets and exciplexes, and in photoconductivity of molecular solids and liquid solutions of aromatic molecules. Later, in 1972, the first reliable and metrologically well-founded MFEs in chemical reactions were discovered [11].

The modern theory of MFEs is based on the extended spin dynamics of the pairs of paramagnetic species (radicals, triplet molecules, etc.) and emphasizes three main sources of MFEs, which will be illustrated below with the classical example of a radical pair.

1. In zero and very low magnetic fields the singlet and triplet spin states of RP are almost degenerated and the spin dynamics involves all four states into the spin conversion (Fig. 3.2), while in high fields the degeneracy of triplet levels is lifted, the $T_{\pm}$ levels are excluded from the conversion, and triplet–singlet fluxes are reduced by approximately two-thirds compared with those in zero field (Fig. 3.2).

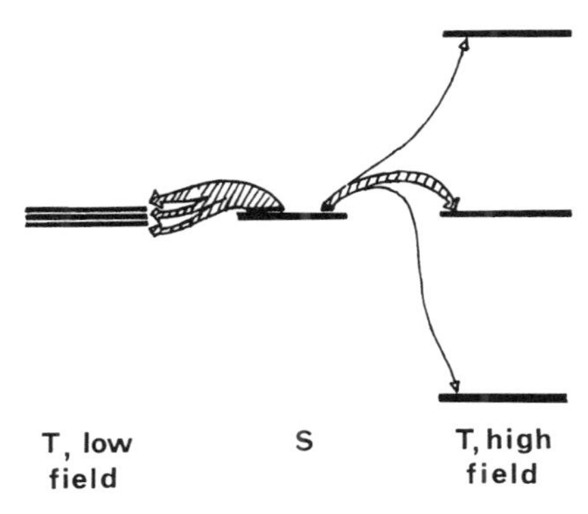

Figure 3.2 Schematic presentation of triplet–singlet fluxes in low (left) and high (right) magnetic fields. The arrow thickness corresponds to the intensity of fluxes. This picture is shown for the initially populated singlet state; the inversion in population results in inversion of the fluxes.

If the starting state of the pair is singlet, the magnetic field destroys the $S \rightarrow T_{\pm}$ channels and partly locks the pair, increasing the competitive chance for pair to react. However, if the starting state is triplet the magnetic field suppresses $T_{\pm} \rightarrow S$ conversion, partly prevents the chemical reaction, and decreases the yield of cage products.

The $T_{\pm} \rightarrow S$ spin conversion is induced by HFI, which usually lies in the range 1–150 G, therefore, even these low magnetic fields $H \approx a$ are sufficient to decouple the electron–nuclear spin subsystems and produce quite noticeable effects in chemistry.

2. The rate of T_0–S spin conversion, as established in Chapter 2, is proportional to $\Delta g\beta H$ and, consequently, depends on the magnetic field. As in the previous case, the sign of the MFE is a function on the starting spin state: the magnetic field enhances the leakage of the reactive singlet pairs into the nonreactive triplet ones, while it results in the opposite effect for the triplet pairs. As a result, the magnetic field accelerates the triplet pair reactions and hinders the reactions in singlets. It is evident that the signs of the MFEs arising from these two sources are opposite. This is an important criterion to identify the spin states of the reactive species via the sign of MFE.

These two sources of MFEs are easily discriminated because they exhibit the different behavior of the MFE magnitude: the magnetic field dependence for the MFEs originating from $T_{\pm}$–S conversion achieves saturation at $H \geqslant a$, while the magnitude of the MFEs induced by T_0–S conversion depends on magnetic field monotonicly. For RP with short-lived radicals ($\tau \ll |Q_{ab}|^{-1}$) the magnitude of the effect is proportional to $(\Delta g\beta H)^2$; for long-lived radical pairs ($\tau \gg |Q_{ab}|^{-1}$) the effect is proportional to $(\Delta g\beta H)^{1/2}$ [12, 13].

3. The third generator of MFEs is the magnetic field dependence of the triplet–singlet conversion rate due to the leakage via other, noncoherent channels of spin conversion. The predominant contribution into the noncoher-

ent leakage goes from the spin-orbit and dipolar interactions, whose field dependence manifests itself in the field dependence of electron relaxation time $T_1(H)$ and, consequently, of the singlet–triplet conversion rate. The first explanation of the magnetic spin effects in radiation chemistry was given by Brocklehurst [14] in terms of this $T_1(H)$ mechanism. It is now obvious that this suggestion is not adequate but deserves respect, being one of the first physically meaningful ideas in spin chemistry.

At present, MFEs are discovered in luminescence and electric conductivity of molecular solids and solutions induced by photolysis and radiolysis, in photochemical and radiation chemical processes in homogeneous and micellar solutions as well as in zeolites, in thermal reactions of decomposition, isomerization, and electron transfer, in reactions of elementoorganic compounds, etc. A profound survey of these effects was given in numerous reviews (e.g. [15, 19]).

Two examples illustrate the mechanisms and magnitude of MFEs. The reactions of fluorosubstituted benzyl chlorides with butyllithium, investigated in Sagdeev et al. [11], are described by the scheme

$$R_1Cl + LiR_2 \rightarrow LiCl + (\dot{R}_1\ \dot{R}_2)^S \begin{cases} \rightarrow R_1R_2 \text{ (recombination)} \\ \rightarrow \dot{R}_1 + \dot{R}_2 \text{ (escape)} \end{cases}$$

where $\dot{R}_1$ are the radicals $C_6F_5CH_2$, p-$FC_6H_4CH_2$, and $(C_6F_5)_2CH$, and $\dot{R}_2$ is the radical C_4H_9.

The precursor of the products of cross-recombination R_1R_2 (cage product) and of symmetric recombination R_1R_1 and R_2R_2 is the singlet RP ($\dot{R}_1\ \dot{R}_2$). The ratio of the product yields $(R_1R_2)/(R_1R_1)$ is a function of the magnetic field (Fig. 3.3). The experimentally detected MFE is in excellent agreement with the

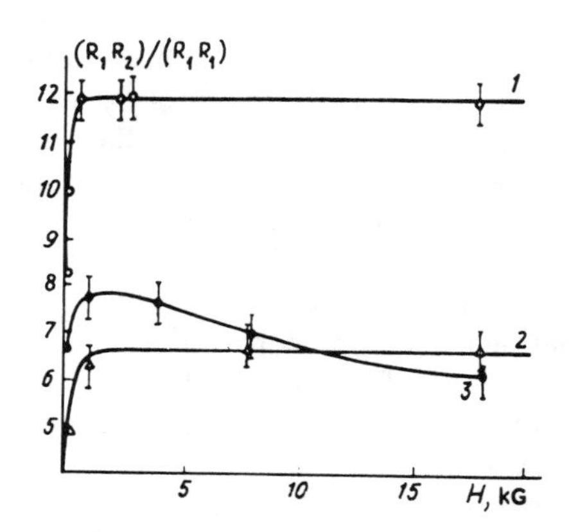

Figure 3.3 Product yield ratio $(R_1R_2)/(R_1R_1)$ as a function of magnetic field in the reaction of C_4H_9Cl with pentafluorobenzyl chloride (1), fluorobenzyl chloride (2), and decafluorodiphenylchloromethane (3).

predictions of the spin theory: the magnetic field kills the $S \rightarrow T_\pm$ conversion channels, partly prevents the leakage of the RP into the nonreactive triplet state, and increases the yield of the cage product. For the first two reactions with $\dot{R}_1 = C_6F_5CH_2$, p-$FC_6H_4CH_2$ in the pair $(\dot{R}_1\ \dot{R}_2)$ $\Delta g \simeq 0$ and the singlet–triplet conversion is induced only by HFI, so that the MFE is saturated even in low fields $H \approx a$; the effect is much greater for the second reaction because the HFI in the radical $C_6F_5CH_2$ (five magnetic fluorine atoms) is approximately five times larger than that in the radical p-$FC_6H_4CH_2$ (single fluorine atom).

In the third reaction with $\dot{R}_1 = (C_6F_5)_2CH$ the low magnetic fields suppress the two $S \rightarrow T_\pm$ channels and result in pair recombination probability enhancement. This is why the ratio $(R_1R_2)/(R_1R_1)$ initially increases. However, in this pair $\Delta g \neq 0$ and the rise of magnetic field is accompanied by the increase in the $S \rightarrow T_0$ conversion rate, which stimulates the leakage of reactive singlet pair into the nonreactive triplets and reduces the recombination probability. As a result of the competition between these two contributions into the singlet–triplet conversion the ratio $(R_1R_2)/(R_1R_1)$ passes through the maximum as the magnetic field increases.

The magnetic field intervention in spin conversion is particularly prominent in chain reactions where the initiation and termination of kinetic chains occur in RP. For example, in dibenzylketone photolysis the excited triplet ketone molecule dissociates into the triplet pair

$$(PhCH_2)_2CO \rightarrow (Ph\dot{C}H_2\dot{C}OCH_2Ph)^T \longrightarrow \begin{cases} PhCH_2COCH_2Ph \\ Ph\dot{C}H_2 + PhCH_2\dot{C}O \end{cases}$$

which then either recombines (after T–S conversion) or dissociates into isolated radicals. The quantum yield of the radicals depends on the magnetic field: it increases by 20% even in low fields $H \approx 100\,\mathrm{G}$ due to the switching off the $T_\pm$–S conversion. For this reason in the chain radical polymerization of styrene in aqueous emulsion initiated by dibenzylketone photolysis the magnetic field increases both the yield of radicals and the rate of initiation of polymerization resulting in the growth of the kinetic chains and the size of the macromolecule. Since the polymerization occurs within the microreactor, the emulsion "droplet" of the monomer, the initial triplet RP whose partners start polymerization, retains its spin correlation. Moreover, the magnetic field additionally stabilizes the triplet pair and prevents its recombination reducing the chain termination rate. Both effects, the higher rate of initiation and the lower termination rate, result in a fivefold increase of the polymerization rate and comparable increase of the macromolecule mass even in low magnetic fields $H \approx 500\,\mathrm{G}$ (Fig. 3.4) [20]. This is an impressive example of how even weak MFEs can be "multiplied up" in chain reactions and reach large magnitudes.

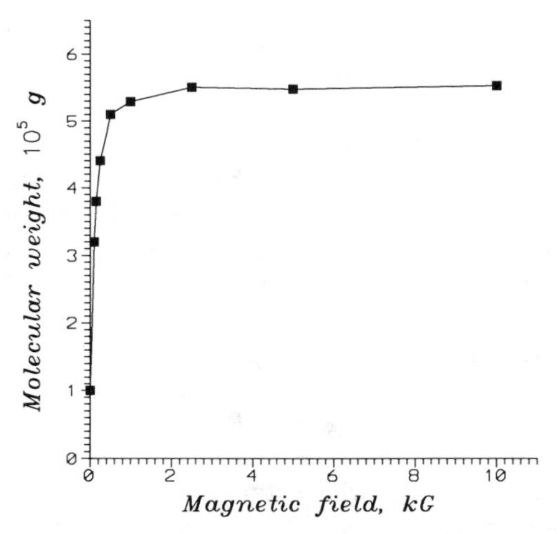

Figure 3.4 The molecular weight of polystyrene as a function of magnetic field in the photopolymerization of styrene; SDS micelles, dibenzylketone as an initiator [20].

3.2. Magnetic Isotope Effect

Chemical interaction between radicals, being electron spin selective, is therefore nuclear spin selective, since the two spin subsystems, electron and nuclear ones, are coupled by the HFI. The selectivity of reactions with respect to spins and magnetic moments of nuclei (i.e., the dependence of the reaction rate on the nuclear magnetic moments of the radicals) results in the difference of the reaction rates of radicals with magnetic and nonmagnetic nuclei. This new phenomenon is the nuclear spin, or *magnetic isotope effect.* It was experimentally discovered in a number of reactions—thermal and photochemical fragmentation of benzoyl peroxide, photolysis of ketones in solutions and in micelles, and chain oxidation of hydrocarbons and polymers.

Its remarkable consequence is the chemically induced separation of magnetic and nonmagnetic isotopes. We will now exemplify the isotope selection by the photolysis of dibenzylketone, which is known to occur via the fragmentation of the excited molecule in the triplet state with the generation of triplet RP. Further, it either undergoes the triplet–singlet conversion and recombines, regenerating the starting ketone, or dissociates into separate radicals $PhCH_2$ and $PhCH_2CO$; the latter decomposes into $PhCH_2$ and CO. The main products of the photolysis are dibenzyl $(PhCH_2)_2$ and carbon monoxide (Scheme 3.1).

The triplet–singlet conversion rate for the magnetic RP (i.e., the pairs with magnetic ^{13}C isotope in benzoyl or benzyl radicals) is much higher than that for nonmagnetic pairs, with nonmagnetic ^{12}C nuclei. Therefore, the magnetic RP has the ability to recombine and to regenerate the starting ketone molecule,

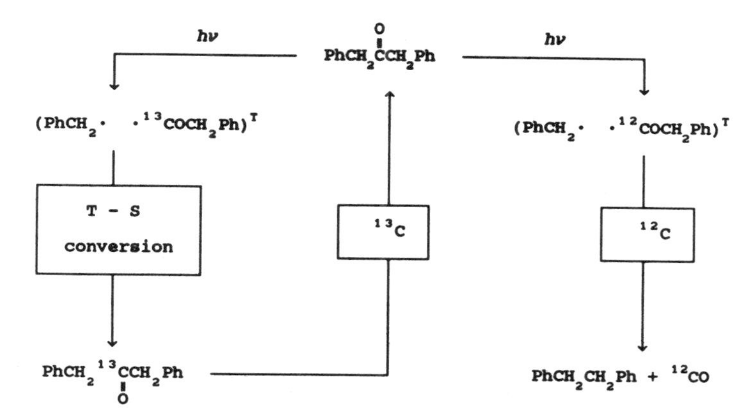

Scheme 3.1 The isotope selection in dibenzylketone photolysis.

while the delay of the triplet–singlet conversion in nonmagnetic RP causes them predominantly to dissociate. As a result, ^{13}C magnetic nuclei are accumulated in the starting ketone and carbon monoxide is enriched with the ^{12}C nonmagnetic isotope.

Thus, due to the difference in the triplet–singlet conversion rates the RP sorts the nuclei according to their magnetic moments: it dispatches the magnetic and nonmagnetic nuclei into different chemical products.

The efficiency of isotope sorting increases as the HFI energy enlarges, the result of the greater difference between the triplet–singlet conversion rates of magnetic and nonmagnetic pairs. Thus, in dibenzylketone photolysis the highest enrichment is inherent to the central carbon atom of the CO group, because the HFI for it in the $PhCH_2CO$ radical is largest, 125 G. For other carbon nuclei of this radical, as well as for those of the partner radical $PhCH_2$, the HFI is smaller and its contribution into isotope sorting is of little importance.

Magnetic isotope effect (MIE) was discovered in 1976–1977 [21, 22]. Turro was the first to show that the MIE magnitude and nuclear isotope selectivity is much higher for the reactions in microreactors, such as micelles, zeolite, and porous glass cavities [23, 24]. Tarasov and Buchachenko [25, 26] explained the high efficiency of MIE-induced isotope separation in microreactors in terms of the high reversibility of the reaction.

Buchachenko has shown that for highly reversible reactions (dibenzylketone photolysis belongs to this type) the isotope enrichment S is related to the chemical conversion as

$$S = (1 - f^*)^{\alpha^*} \qquad (3.1)$$

where f^* is the chemical conversion measured with respect to [^{13}C]ketone, or

$$S = (1 - f)^{\alpha} \qquad (3.2)$$

where f is the chemical conversion of $[^{12}C]$ketone. In general

$$S = (1 - f_{expt\ 1})^{\alpha}[(1 + \delta_0)/(1 + \delta)]^{\alpha} \tag{3.3}$$

where $f_{expt\ 1}$ is the total chemical conversion of ketone, δ_0 and δ are the ^{13}C isotope content in the starting reagent and in ketone that remains after the conversion $f_{expt\ 1}$ is achieved, and $S = \delta/\delta_0$ (see Buchachenko [26] for details).

The isotope enrichment depends strongly on conversion: the higher the latter the stronger the isotope selection and the isotope enrichment of the remaining ketone. The value α (or α^*) is an important parameter that controls isotope enrichment.

According to the ketone A decomposition kinetic scheme

$$A \underset{k_2}{\overset{w_i}{\rightleftarrows}} RP \xrightarrow{k_1} \text{escape product}$$

where w_i is the rate of RP generation and the value α is defined by the equation

$$\alpha = \frac{1 - \gamma}{(k_2/k_1)^{-1} + \gamma}, \tag{3.4}$$

where two parameters, $\gamma = k_2^*/k_2$, the ratio of recombination rates of magnetic and nonmagnetic pairs, and k_2/k_1, the ratio that characterizes the reaction reversibility, are of prime importance.

On the other hand, α and α^* can be expressed via the recombination probability of magnetic, P^*, and nonmagnetic, P, radical pairs:

$$\alpha = (P - P^*)/(1 - P), \qquad \alpha^* = (P - P^*)/(1 - P^*) \tag{3.5}$$

The relations (3.1)–(3.5) allow the extraction of the kinetic parameters of the reaction and values of P and P^* from the enrichment experimentally measured as a function of the conversion. Their values can be calculated theoretically within the scope of the extended spin dynamics (Chapter 2) and compared with experimental data [19, 26].

The MIE has been already discovered for carbon [26], hydrogen [27], oxygen [28, 29], sulfur [30], silicon [31] isotopes, and, finally, for the isotope pair ^{235}U–^{238}U [32].

Three remarkable features distinguish the MIE from the standard, classical isotope effect induced by the difference in the masses of isotope nuclei.

1. The magnitude of the MIE is much higher than that of the classical isotope effect and, moreover, in contrast to the latter, actually has no upper limit.
2. Unlike the classical effect, MIE is magnetic field dependent: it drops as the magnetic field increases due to the suppression of HFI-induced $T_{\pm}$–S conversion, which makes the main contribution into the magnetic isotope selection.
3. The MIE depends on the HFI energy, on the spin and magnetic moments of nuclei, as well as on the parameters characterizing molecular dynamics (viscosity, radical diffusion coefficients) and chemical dynamics (the lifetimes of radicals and radical pairs).

MIE is a phenomenon of fundamental importance. It is the basis for a new principle of isotope and isomer nuclei separation, it furnishes a new method of paramagnetic intermediates detection in chemical and biochemical processes, and offers new ideas and approaches to many problems of geochemistry, space chemistry, biophysics, and molecular biology [26].

3.3. Chemically Induced Dynamic Nuclear Polarization

One remarkable property of the chemical reaction is the chemically induced dynamic nuclear polarization (CIDNP) discovered by Bargon, Fischer, Lawler, and Ward in 1967. As shown in Chapter 2, the rate of T_0–S conversion in RP depends on the nuclear spin orientation and, consequently, RP *with different nuclear spin orientations exhibit different chemical reactivity* as well. Hence, the RP not only sorts magnetic and nonmagnetic nuclei (as it does in MIE), but also distinguishes the orientations of the magnetic nuclei: nuclei with different spin orientations are distributed among different products.

This is the first mechanism of CIDNP and it works in high magnetic fields, when the S–T_0 channel of spin conversion is predominant. It produces spin selection in accordance to Eqs. (2.1), (2.5), and (2.12) with the rate proportional to $\Delta g\beta H$ and a, hence, the magnitude of CIDNP is proportional to Δg in RP and to the HFI constants.

The second CIDNP mechanism operates in low fields $H < a$ when S–$T_{\pm}$ channels also participate in spin evolution. According to Eqs. (2.2), (2.3), (2.6), and (2.7) they are accompanied by the nuclear spin reorientation, $T_+\beta_n \rightarrow S\alpha_n$ or $T_-\alpha_n \rightarrow S\beta_n$. As a rule, the exchange interaction in RP is negative; it aligns the electron spins in an antiferromagnetic manner, hence, the pair ground state is singlet. It means that the T_- level is closer to the S level than T_+ one, and, consequently, the $T_-\alpha_n \rightarrow S\beta_n$ transition rate exceeds that of the $T_+\beta_n \rightarrow S\alpha_n$ transition. If the starting spin state of RP is triplet, then the former results in an excess of nuclear spins β_n in all reaction products. If the starting state of RP is singlet, then the predominance of $S\beta_n \rightarrow T_-\alpha_n$ transitions results in the excess of nuclear spins α_n in the products.

The identical sign of nuclear polarization for the different reaction products is a specific symptom of the second mechanism of CIDNP. It appears predominantly in the low fields, where Zeeman energy is comparable with exchange energy, and, therefore, even a weak exchange interaction differentiates the relative contribution of T_+ and T_- levels into $T_{\pm}$–S conversion. Usually, this mechanism prevails in micelles or in biradicals (i.e., in molecular systems where the exchange interaction cannot be neglected).

The excess of α_n or β_n nuclear spins in the product molecules corresponds to the overpopulation of the lower or upper Zeeman nuclear levels, respectively. In the NMR spectra of the molecules this nonequilibrium population manifests itself as anomalously strong absorption or emission; the latter is the case when the populations of Zeeman levels are inverted. The former case corresponds to the positive and the latter case to the negative nuclear polarization.

Now we will present some illustrations of the CIDNP phenomenon. The thermal decomposition of benzoylperoxide generates benzene molecules, one of the main decomposition products, whose protons are polarized negatively (Fig. 3.5). The intensity of the NMR emission line first increases, goes through a maximum, and then, after 10–12 min, reaches an ultimate value corresponding to the signal of normal NMR absorption of benzene molecules.

The ^{13}C NMR spectra of acetylbenzoylperoxide thermal decomposition products demonstrate two types of ^{13}C nuclear polarization—both positive, for $PhCH_3$ and CH_3OCOPh, and negative, for CH_4 and CH_3Cl (Fig. 3.6). The former are the products of cage recombination of methyl radicals with the phenyl and benzoyloxy radical, and the latter are formed from the methyl radicals that avoid the cage recombination. In accordance with the predictions of CIDNP theory they exhibit the polarization of opposite signs.

Figure 3.7 shows the NMR spectra of ethyl- and 2-iodobutane protons arising in reactions of these molecules with organolithium compounds. Unusual multiplet polarization of the ethyl- and isobuthyl groups is produced in reactions

$$R_1I + LiR_2 \rightarrow (\dot{R}_1\ \dot{R}_2) + LiI$$

which generates RP $(\dot{R}_1\ \dot{R}_2)$, where $R_1 = C_2H_5$, $CH(CH_3)CH_2CH_3$, $R_2 = CH_2CH_2CH_2CH_3$. The selection of radicals according to their nuclear spin orientations in these pairs produces unusual nuclear polarization in these radicals, which is transferred to the alkyliodides as a result of the exchange

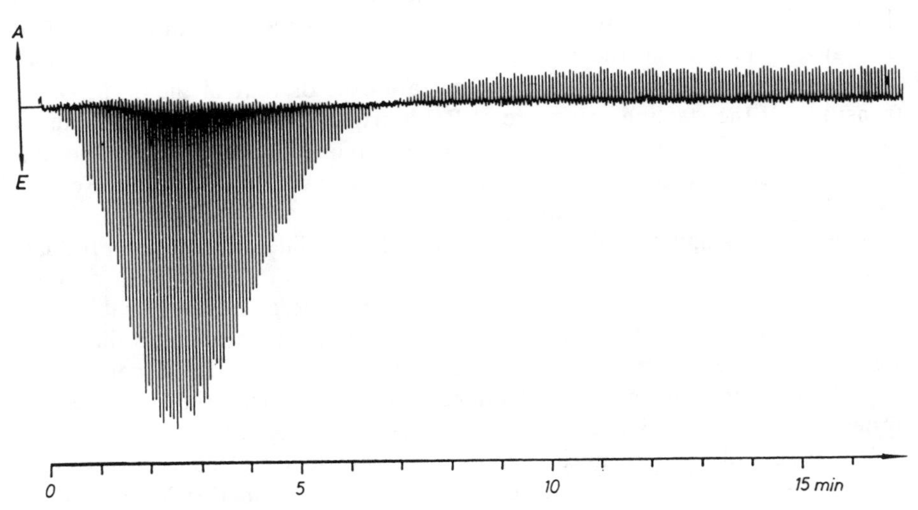

Figure 3.5 Time behavior of the proton NMR signal of benzene in the benzoyl peroxide thermolysis (110°C, 5 mol% of peroxide in cyclohexanone) [33].

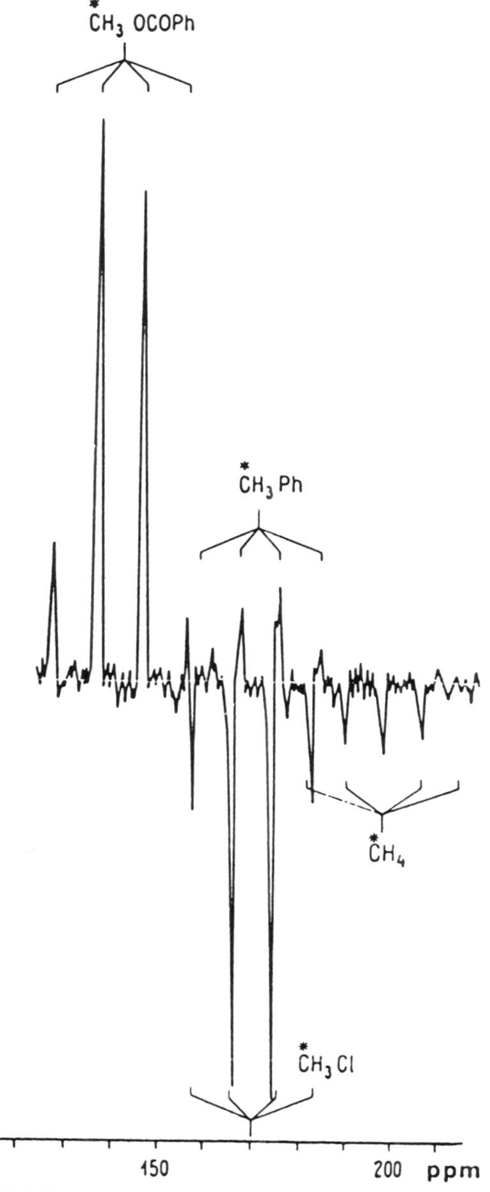

Figure 3.6 ^{13}C CIDNP spectrum for the products of acetylbenzoyl peroxide thermolysis in C_2Cl_4 at 110°C. Polarized nuclei are marked by asterisks [34].

reactions $\dot{R} + RI \rightarrow R + \dot{R}I$, where $\dot{R}$ designates ethyl or isobuthyl species with the polarized nuclei.

Finally, the ^{19}F NMR spectrum of pentafluorobenzene generated by thermolysis of di(pentafluorobenzoyl)peroxide is depicted in Fig. 3.8. The total

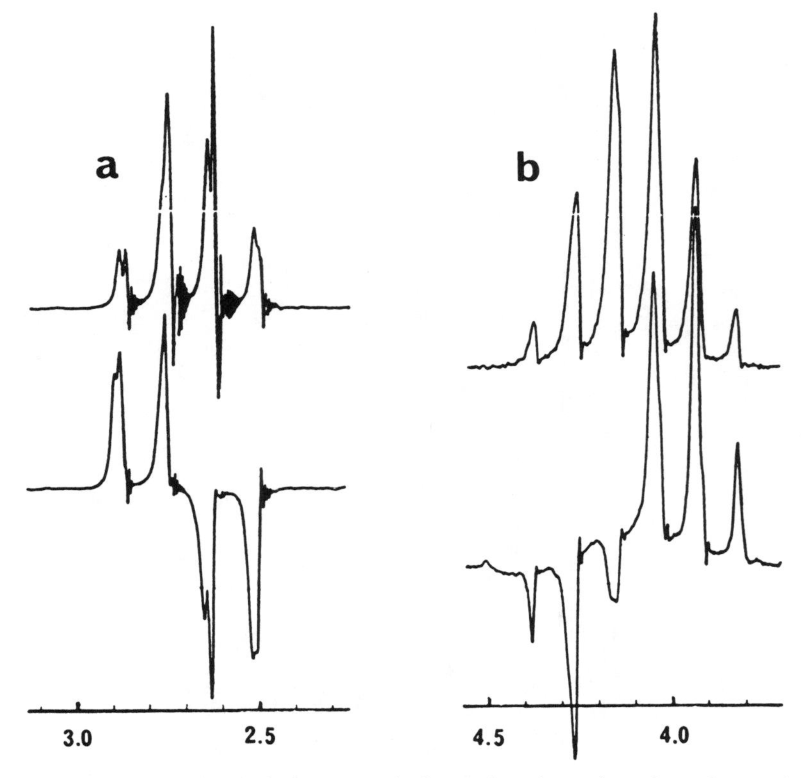

Figure 3.7 A comparison of the normal absorption (upper) and nuclear polarized (lower) proton magnetic resonance lines for the methylene group (a) of ethyliodide and CH group (b) of 2-iodobutane. Polarization is generated in reactions of alkyl iodides with organolithium compounds and transferred in alkyliodide molecules via the exchange reactions [35].

inversion of the spectrum corresponds to the negative polarization of ^{19}F nuclei in the ortho position of this molecule.

In the quarter of a century since its development CIDNP has become the most efficient method for the identification of chemical reaction mechanisms and the detection of radicals and radical intermediates. By means of CIDNP, it is easy to define the spin multiplicity of pairs and to determine the states—singlet or triplet—from which radicals and molecules are generated. The method permits separation of radical and nonradical pathways of the reactions, to identify unstable intermediate products and reversible radical stages that fail to be determined by other methods. From the CIDNP kinetics one can determine the reaction rate constants including that of the fast reactions in RP (decomposition, radical isomerization, substitution reactions, electron transfer, etc.), which develop for characteristic times of 10^{-8}–10^{-9} s.

CIDNP is a powerful tool to determine the signs of HFI constants in

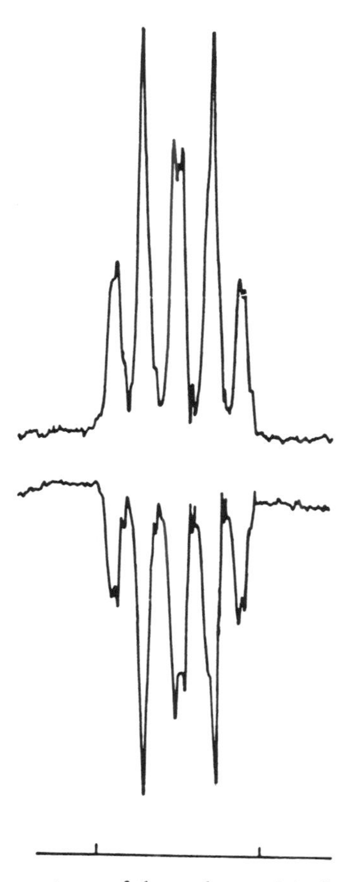

Figure 3.8 The ^{19}F NMR spectrum of the ortho nuclei of pentafluorobenzene (upper) and the spectrum of the same molecule (lower) generated by thermolysis of di(pentafluorobenzoyl) peroxide. The spectrum covers 1 ppm [36].

radicals, the signs of spin–spin coupling constants in molecules, and the nuclear relaxation times in the radicals, and to detect the hot, energy-rich radicals. By means of CIDNP the mechanisms of numerous types of chemical reactions are investigated and much new information has been collected and pooled in many books and reviews (see, for instance [13, 19, 37–40]).

3.4. Chemically Induced Dynamic Electron Polarization

This spin effect is caused by the nonequilibrium population of electron Zeeman levels in radicals and manifests itself in their ESR spectra, either as an abnormally strong absorption when the lower Zeeman level is overpopulated, or as an emission when the upper Zeeman level is inversely populated. Two physical mechanisms are responsible for the nonequilibrium pumping of the electron Zeeman levels—the first active in RP and the second in triplet molecules.

The spin dynamics, as shown in Chapter 2, is selective with respect to the electron and nuclear spins and results in the recombination of RP in selectively populated nuclear spin states. The population of electron and nuclear spin states in the radicals, which avoid the recombination, is also nonequilibrium and is almost a mirror reflection of the nonequilibrium population of the recombined radicals. Thus, the hydrogen atom that leaves RP carries the α electron spin preferentially with the α_n nuclear spin, while the β electron spin merges the β_n. The preferable population of the spin states $\alpha\alpha_n$ and $\beta\beta_n$ results in an unusual ESR spectrum of the hydrogen atoms shown in Fig. 3.9.

However, in contrast to CIDNP, the chemically induced dynamic electron polarization (CIDEP) requires both spin mixing and exchange $J(r)$ interaction. No electron polarization results from the simple spin evolution since the S and T_0 states contain an equal amount of α and β spin character on each radical. CIDEP owes its origin not merely to a spin-sorting process, as CIDNP does, but rather to the action of the J interaction producing a phase shift in the spin-mixed electronic wave function of RP. The theory of this phenomenon is more complex and widely reviewed [19, 42–44].

Strictly speaking, the CIDEP generated according to the second mechanism is not actually chemically induced. It is produced in triplet molecules due to the fact that the singlet–triplet transitions in the excited molecules populate the T_x, T_y, T_z sublevels (and, therefore, T_+, T_0, T_-) with different rates because of the spin–orbital coupling anisotropy. In addition, the rates of depopulation of these sublevels are also different due to dipolar interaction anisotropy. It causes nonequilibrium populations of the substates $T_\pm$ and T_0 of the triplet molecule. If such a molecule reacts its nonequilibrium population is transferred to radicals and is detected in their ESR spectra. In other words, the chemical

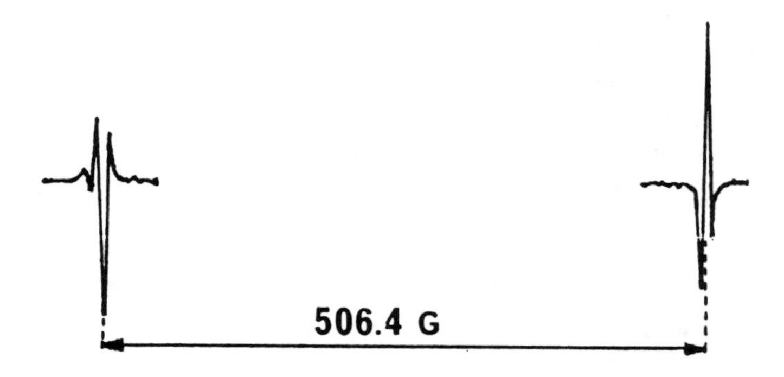

Figure 3.9 Electron spin resonance spectrum (second derivative) of hydrogen atoms showing CIDEP effect. Low field line in emission, high field line in absorption (EA multiplet effect) [41].

reaction fixes only the electron polarization created in the molecule, the radical precursor. For the electron polarization to be fixed by chemical reaction the lifetime of the triplet molecule must be shorter than the spin–lattice relaxation time. The molecule with a nonequilibrium population of triplet sublevels should have a chance to transfer it into the radicals before the spin–lattice relaxation occurs.

The characteristic features of CIDEP produced by the triplet mechanism are as follows. First, both partner radicals have a polarization identical in sign and magnitude. Second, the magnitude of the polarization depends on neither the HFI energy nor the g-factors of radicals. Third, the polarization decays through the spin–lattice electron relaxation in radicals. Taking into account these features, one can easily distinguish between the triplet born CIDEP and that generated in radical pairs.

References

1. Bhatagnar, S. S., Mathur, R. N., Kapur, R. N. *Philos. Mag.* **1929**, *8*, 457.
2. Figueras Roca, F. *Ann. Chim.* **1967**, *2*, 255.
3. Atkins, P. W., Lambert, T. P. *Ann. Rep. Prog. Chem.* **1975**, *72A*, 67.
4. Frankevich, E. L., Balabanov, E. I. *JETP Lett.* **1965**, *1*, 169.
5. Frankevich, E. L., Balabanov, E. I., Vselyubskaya, G. V. *Sov. Phys.-Solid State* (Engl. Transl.) **1966**, *8*, 1567.
6. Merrifield, R. E., Avakian, P., Groff, R. P. *Chem. Phys. Lett.* **1969**, *3*, 155.
7. Geacintov, N. E., Pope, M., Vogel, F. *Phys. Rev. Lett.* **1969**, *22*, 593.
8. Pope, M., Geacintov, N. E., Vogel, F. *Mol. Cryst. Liq. Cryst.* **1969**, *6*, 83.
9. Johnson, R. C., Merrifield, R. E. *Phys. Rev. B* **1970**, *1*, 896.
10. Geacintov, N. E., Pope, M., Fox, S. *J. Phys. Chem. Solids* **1970**, *31*, 1375.
11. Sagdeev, R. Z., Salikhov, K. M., Leshina, T. V., Kamkha, M. A., Shein, S. M., Molin, Yu. N. *JETP Lett.* **1972**, *16*, 422.
12. Buchachenko, A. L., Markaryan, S. A. *React. Kinet. Catal. Lett.* **1974**, *1*, 157.
13. Buchachenko, A. L. *Chemically Induced Magnetic Polarization of Electrons and Nuclei.* Science, Moscow, 1974.
14. Brocklehurst, B. *Nature (London)* **1969**, *221*, 921.
15. Swenberg, Ch. E., Geacintov, N. E. Magnetic field effects in luminescence. In *Organic Molecular Photophysics*, Birks, J. B., ed. Wiley, New York, 1973.
16. Sokolik, I. A., Frankevich, E. L. *Sov. Phys. Rev.* **1973**, *111*, 261.
17. Salikhov, K., Molin, Yu., Sagdeev, R., Buchachenko, A. *Spin Polarization and Magnetic Effects in Radical Reactions.* Elsevier, Amsterdam, 1984.
18. Sagdeev, R. Z., Salikhov, K. M., Molin, Yu. N. *Russ. Chem. Rev.* **1977**, *46*, 297.
19. Steiner, U. E., Ulrich, T. *Chem. Rev.* **1989**, *89*, 51.

20. Turro, N. J., Chow, M.-F., Chung, Ch.-J., Tung, Ch.-H. *J. Am. Chem. Soc.* **1983**, *105*, 1572.

21. Buchachenko, A. L., Galimov, E. M., Ershov, V. V., Nikiforov, G. A. *Dokl. Acad. Nauk SSSR* **1976**, *228*, 379.

22. Sagdeev, R. Z., Leshina, T. V., Kamkha, M. A., Belchenko, O. I., Molin, Yu. N., Rezvukhin, A. J. *Chem. Phys. Lett.* **1977**, *48*, 89.

23. Turro, N. J., Krautler, B. *J. Am. Chem. Soc.* **1978**, *100*, 7432.

24. Turro, N. J., Cheng, C.-C., Wan, P., Chung, C-J., Mahler, W. *J. Phys. Chem.* **1985**, *89*, 1567.

25. Tarasov, V. F., Buchachenko, A. L. *Russ. J. Phys. Chem.* **1981**, *55*, 1921.

26. Buchachenko, A. L. Magnetic isotope effect and isotope selection in chemical reactions. In *Progress in Reaction Kinetics*, Jennings, K. R., Cundall, R. B., Margerum, D. W., eds. Pergamon Press, Oxford, 1984, Vol. 13, No. 3, pp. 164.

27. Buchachenko, A. L., *Proc. Sov. Acad. Sci. Chem. Ser.* (Engl. Transl.) **1990**, 2045.

28. Buchachenko, A. L., Fedorov, A. V., Yasina, L. L. *Chem. Phys. Lett.* **1984**, *103*, 405.

29. Turro, N. J., Chow, M.-F., Rigaudy, J. *J. Am. Chem. Soc.* **1981**, *103*, 7218.

30. Step, E. N., Tarasov, V. F., Buchachenko, A. L. *Nature (London)* **1990**, *345*, 17.

31. Yasina, L. L., Buchachenko, A. L. *Chem. Phys.* **1990**, *146*, 225.

32. Buchachenko, A. L., Khudyakov, I. V. *Acc. Chem. Res.* **1991**, *24*, 177.

33. Bargon, J., Fischer, H., Johnsen, U. *Zeit. Naturforsch.* **1967**, *22a*, 1551.

34. Lippmaa, E. T., Pehk, T. I., Buchachenko, A. L., Rykov, S. V. *Chem. Phys. Lett.* **1970**, *5*, 521.

35. Lepley, A. R. *J. Chem. Soc. Chem. Commun.* **1969**, 64.

36. Roth, H., Kaplan, M. *J. Am. Chem. Soc.* **1973**, *95*, 262.

37. Lepley, A. R., Closs, G. L., eds. *Chemically Induced Magnetic Polarization.* Wiley, London, 1978.

38. Roth, H. D. *Mol. Photochem.* **1973**, *5*, 91.

39. Ward, H., Lawler, R., Rosenfeld, S. *Ann. N.Y. Acad. Sci.* **1973**, *222*, 740.

40. Buchachenko, A. L., Zhidomirov, G. M. *Russ. Chem. Rev.* (Engl. Transl.) **1971**, *40*, 801.

41. Eiben, K., Fessenden, R. *J. Phys. Chem.* **1971**, *75*, 1186.

42. McLauchlan, K. A., Steiner, U. E. *Mol. Phys.* **1991**, *73*, 241.

43. McLauchlan, K. A. Chemically induced dynamic electron polarization. In *Modern Pulse and Continuous-Wave Electron Spin Resonance,* Kevan, L., Bowman, M., eds. Wiley, London, 1990.

44. McLauchlan, K. A. In *Applications of Lasers in Polymer Science and Technology*, Fonassier, J. F., Rabek, J. F., eds. CRC Press, London, 1990.

CHAPTER

4

Chemically Induced Radio-Frequency Emission

Chemically induced polarization of nuclei results in nonequilibrium or even inverted population of the nuclear Zeeman levels in the product molecules. The inverted population corresponds to energy storage in the nuclear Zeeman reservoir of molecules. This new phenomenon brings up new questions: What is the magnitude of the energy that it is possible to pump chemically into the molecular nuclear spin reservoir? Could the energy stored therein be released as the radio-frequency emission? What interactions stimulate the release and under what conditions? The answers to these questions will be discussed in this chapter.

4.1. Chemical Pumping

The magnitude of the nuclear polarization is determined by the coefficient Z:

$$Z = (n_+ - n_-)/(n_+^0 - n_-^0) \qquad (4.1)$$

where n_-^0, n_+^0 and n_-, n_+ are, respectively, the equilibrium and nonequilibrium populations of the upper and lower nuclear Zeeman levels. It shows at what times the nonequilibrium difference of the populations in newly born molecule exceeds its equilibrium value.

The quantity Z depends on the nuclear spin and magnetic moments, on the HFI energy, on the difference between the g-factors of radicals in the radical pair (RP), and on the magnetic field (i.e., on the molecular and magnetic parameters that govern the RP spin dynamics), and can be calculated according to the theory of extended spin dynamics (Chapter 2).

It is not difficult to calculate the energy pumped by chemical reaction and stored in the Zeeman reservoir. The ratio of equilibrium populations is

$$n_+^0/n_-^0 = \exp(\hbar\gamma_n H/2kT) \approx 1 + \hbar\gamma_n H/2kT \tag{4.2}$$

where $\hbar\gamma_n H$ is the radio-frequency "quantum" (i.e., the energy excess of the nucleus on the upper Zeeman level) and γ_n is the gyromagnetic ratio of the nucleus. From this expression the Zeeman energy of N polarized nuclei in the magnetic field H can be derived:

$$E_Z = N(\hbar\gamma_n H)(\hbar\gamma_n H/2kT)|Z| \tag{4.3}$$

For example, in the field of 15 kG Zeeman energy of 1 mol of protons with the polarization coefficient $|Z| = 10^3$ is about 2×10^{-5} J at 30°C.

The energy of the Zeeman reservoir is proportional to the nuclear polarization coefficient, therefore the chemically induced emission is most likely to be found in the reactions with strong nuclear polarization; Figs. 4.1 and 4.2 illustrate such a type of polarization; other examples were given in Chapter 3.

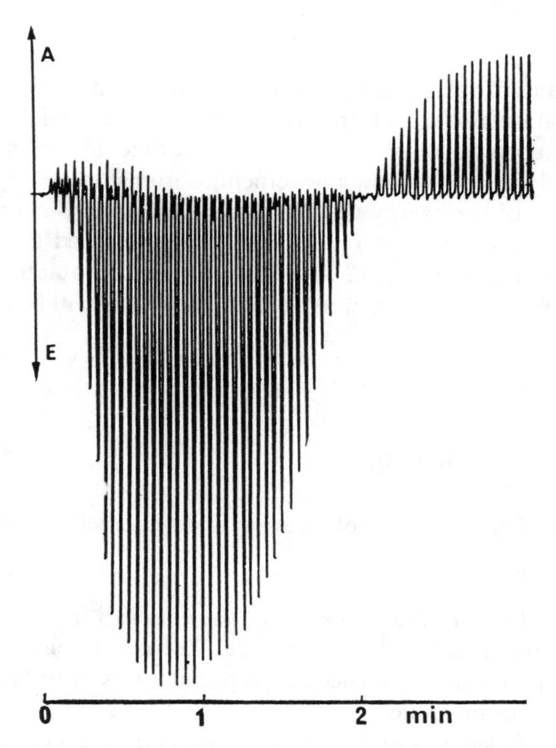

Figure 4.1 The kinetics of the CIDNP signal of benzene molecules polarized in the reaction of phenyldiazonium cation with sodium ethylate [3].

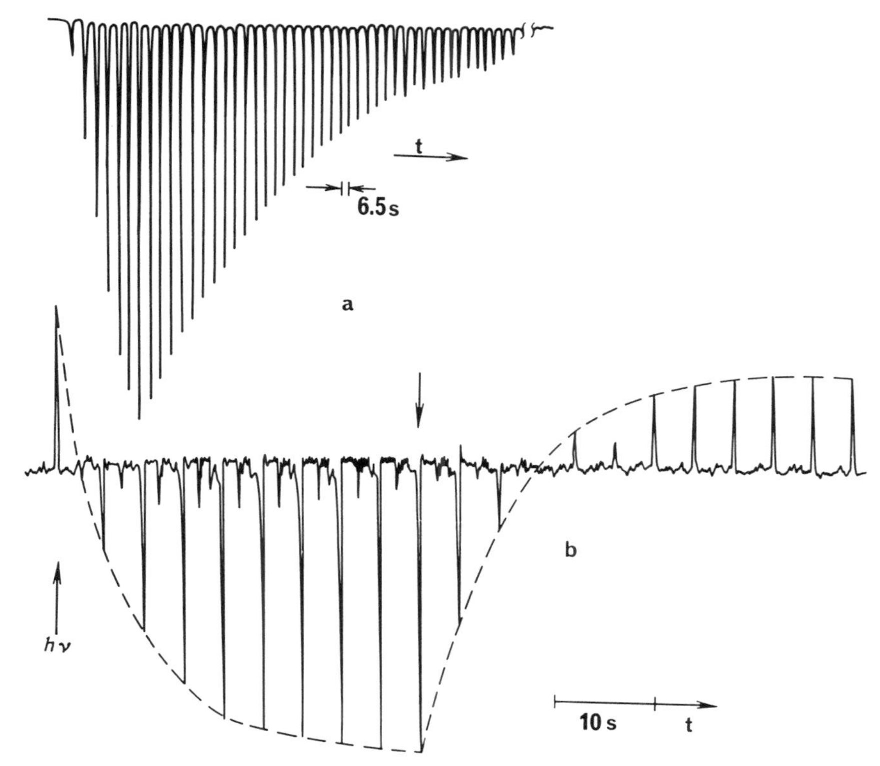

Figure 4.2 CIDNP of ^{31}P nuclei in trimethylphosphate in reaction with cyclohexylperoxydicarbonate (a) and of CH_2 protons of dibenzylketone (b). Arrows indicate the switching on/off of the photolysis [4].

The chemical pumping is a nonstationary process; it strictly follows the chemical reaction kinetics and is described by the general equations [1, 2]:

$$\frac{dM}{dt} = Z\frac{dM_0}{dt} - \beta(M - M_0) \tag{4.4}$$

or

$$\frac{dI}{dt} = Z\frac{dI_0}{dt} - \beta(I - I_0) \tag{4.5}$$

where M is the nuclear magnetization of some sorts of molecules or molecular groups and fragments, M_0 is the equilibrium nuclear magnetization of the same molecules or groups, and I and I_0 are the intensities of the NMR signals, proportional to M and M_0, respectively. The first term of the Eqs. (4.4) and (4.5) determines the rate of chemical pumping; the second term describes the rate of relaxation of the nuclear nonequilibrium magnetization M to its equilibrium value M_0; $\beta = T_{1n}^{-1}$ is the relaxation rate constant and T_{1n} is the nuclear spin-lattice relaxation time.

$$M_0 = \mu_n p_{n0} P, \qquad \frac{dM_0}{dt} = \mu_p p_{n0} \frac{dP}{dt} \tag{4.6}$$

$$p_{n0} = \frac{n_+^0 - n_-^0}{n_+^0 + n_-^0} = \tanh \frac{\hbar \gamma_n H}{2kT}$$

Here μ_n is the magnetic moment of the nucleus and P denotes the number of the product molecules.

The solutions of Eqs. (4.4) and (4.5) for different reaction kinetics have been summarized [1]. For the particular but very important case of unimolecular transformation of the molecules A

$$A \xrightarrow{\kappa} P^*$$

into the product molecules P* with polarized nuclei the solution of the Eq. (4.4) describes the dynamics of the nuclear spin magnetization:

$$M - M_0 = \frac{\mu_n p_{n0} \kappa A_0 Z}{\beta - \kappa} [\exp(-\kappa t) - \exp(-\beta t)] \tag{4.7}$$

$$M_0 = \mu_n p_{n0} A_0 [1 - \exp(-\kappa t)] \tag{4.8}$$

Here A_0 is the initial concentration of A and κ is the reaction rate constant.

For the analysis of nuclear polarization kinetics it is convenient to use the value K:

$$K = \frac{M - M_{0\infty}}{M_{0\infty}} = \frac{I - I_{0\infty}}{I_{0\infty}} \tag{4.9}$$

$$= \frac{Z\kappa - \beta}{\beta - \kappa} [\exp(-\kappa t) - \exp(-\beta t)] - \exp(-\beta t)$$

Here $M_{0\infty}$ and $I_{0\infty}$ are $\lim M_0$ and $\lim I_0$ at $t \to \infty$. Equations (4.7) and (4.9) predict the maximum of nuclear polarization, which is achieved at

$$t_{\max} = \frac{1}{\beta - \kappa} \ln \frac{(Z-1)\beta}{Z\kappa - \beta} \tag{4.10}$$

and is determined by the rates of two competitive processes—chemical pumping and nuclear relaxation.

For a slow reaction ($\kappa \ll \beta$) after $t \gg T_{1n}$ the decay of the nuclear polarization occurs exponentially:

$$K = (Z\kappa T_{1n} - 1) \exp(-\kappa t) \tag{4.11}$$

All these predictions are in agreement with experimental kinetics of chemical pumping (see Figs. 4.1 and 4.2). There is no problem in generalizing the solutions of the Eqs. (4.4) and (4.5) to predict the dynamic behavior of nuclear magnetization for any reaction kinetics.

As noted in Chapter 3, a remarkable feature of CIDNP is the large polarization coefficients $|Z| \approx 10^2$–10^4; moreover reactions were found in which Z achieves the huge values approaching the limiting value

$$Z_{\text{lim}} = (2kT/\hbar\gamma_n H) \tag{4.12}$$

Thus, cyclododecanone protons exhibit anomalously high negative polarization of $|Z| \geqslant 3.6 \times 10^6$ (liquid cyclododecanone is photolyzed in the magnetic field of 125 G), which corresponds to a quantum "yield" of the nuclear polarization of more than 30% (i.e., every two to three photochemically active photons transfer one proton from the lower Zeeman level to the upper one [5]).

In the photosensitized decomposition of cyclohexanone diperoxide in the solution at 20°C in the magnetic field of 200 G, the protons of cyclohexane, the reaction product, are polarized with $Z = -4 \times 10^5$ [6]. Though this value is an order of magnitude lower than that in the previous case, and the polarization quantum yield is only 3.6%, it is still surprisingly large, being far beyond the range accessible by using existing physical methods of nuclear spin alignment (for example, through Overhauser effect).

Clearly, the chemical reactions can generate huge nuclear polarization in reaction products and store the large amount of energy in their Zeeman reservoir. Now the problem is to define the conditions for emissive release of this energy.

4.2. Self-Excitation of Radio-Frequency Generation

The inversion of Zeeman level populations is a necessary but by no means sufficient condition for the emission of energy stored in the Zeeman reservoir. The probability of spontaneous radiative transitions between the nuclear Zeeman levels is very small. Thus, for a proton in the field of 7500 G it is 10^{-25}, which corresponds to an emission time of 10^{17} years [7]. This is much greater than the spin–lattice nuclear relaxation time (1–100 s), during which Zeeman level populations achieve the thermal, Boltzmann, equilibrium, when the energy of the Zeeman reservoir is irreversibly converted into the thermal energy of the lattice. Therefore it is important to establish the conditions under which Zeeman energy of the reaction products is emitted as radio-frequency (RF) radiation instead of being converted and wasted into heat by the spin–lattice relaxation processes.

The magnetic moments of the negatively polarized nuclei μ_n are combined into the total magnetization vector **M** oriented opposite to the magnetic field. For the reaction products to be able to generate RF radiation and to create an electromagnetic field (EMF), the self-excitation of the Larmor precession of the nuclear magnetization vector **M** round the direction of the external field H should take place (Fig. 4.3). The precessing magnetization vector is a source of EMF and of magnetic dipolar radiation.

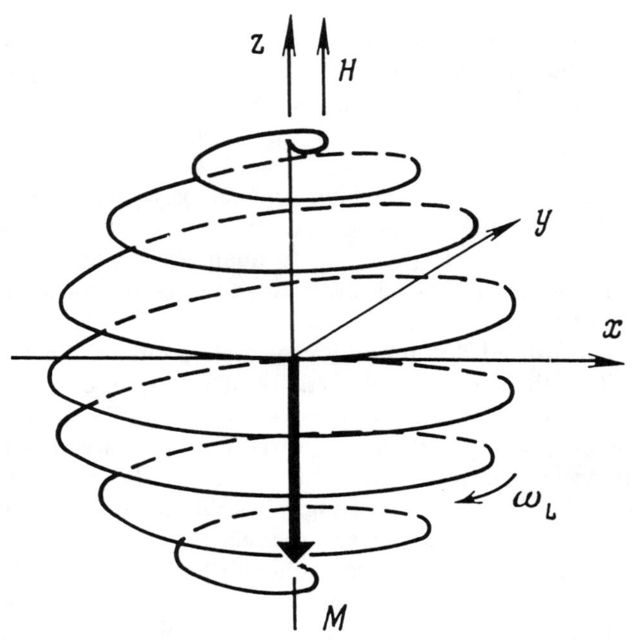

Figure 4.3 Precession of the magnetization vector of negatively polarized nuclei. The transverse components M_x and M_y oscillate in the xy plane with Larmor frequency ω_L.

The self-excitation of the precession **M** means that the ensemble of molecules with polarized nuclei acquires the transverse components of magnetization $M_x(t)$ and $M_y(t)$, oscillating with the Larmor precession frequency $\omega_L = \hbar\gamma_n H$ in the xy plane perpendicular to the direction of H. The transverse component of magnetization $M_\perp = (M_x^2 + M_y^2)^{1/2}$ induces EMF with the frequency of ω_L in the induction coil where a sample with chemically polarized nuclei is placed. This process is identical to the self-excitation of EMF in any electric generator coil, however, the function of the rotating magnet is performed by the transverse component of the precessing total nuclear magnetic moment.

Self-excitation of the precession of the vector **M** means the appearance of spontaneous synchronization (coherence) in the motion of negatively polarized nuclear magnetic moments (Fig. 4.4). In the absence of coherence the individual nuclei oriented against the field H precess independently, their precession phases are random, and the transverse components of their magnetic moments are chaotically "scattered" in the precession plane xy. As a result, there is no net transverse magnetization ($M_\perp = 0$). However, due to the stochastic fluctuations of the local magnetic fields (magnetic noise) the magnetic moment **M** may slightly deviate from the axis z resulting in a small transverse component of the magnetic moment that induces a small EMF in the coil. In turn, this reactive EMF affects the magnetic moments of the nuclei and compels them to

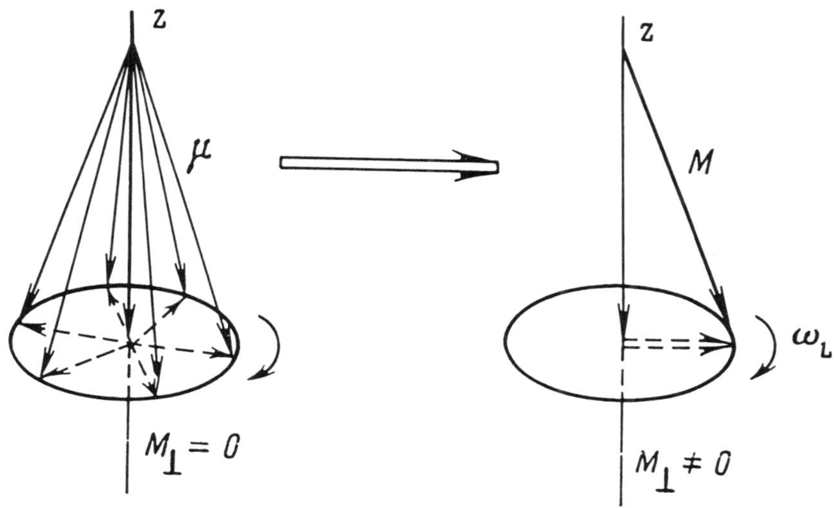

Figure 4.4 Transition from noncoherent motion to coherent precession of negatively polarized nuclear magnetic moments (coherentization process).

precess consistently, synchronously, and coherently, so that all the precessions are in phase, and the transverse components become aggregated in the total moment $M_{\perp} = 0$. In other words, the resulting coherent EMF of the reaction induces a further avalanche-like increase in the synchronization of the precession of the individual nuclear magnetic moments (Fig. 4.4) and stimulates the motion of the total vector **M**, as shown in Fig. 4.3, which creates EMF generation in the electric circuit of the coil. If the negative magnetization M is so large that the energy transferred from the nuclear spin system to the circuit exceeds the electric losses in it, such a system becomes an RF generator whose self-oscillations are supported by continuous chemical pumping.

The coherence is destroyed by the transverse magnetic relaxation processes due to the random local magnetic fields H_{loc}, which cause a scatter in the frequencies and phases of the magnetic moment precession of the individual nuclei, and, as a result, the decay of the transverse components of the magnetization vector. This process is opposite to coherentization (Fig. 4.4). To ensure coherence, the field H^* created by the chemical reaction products, which "collects" the transverse components of the magnetization and synchronizes their precessions, must be greater than H_{loc}, $H^* > H_{\text{loc}}$. In other words, the coherentization time T_2^* must be shorter than the transverse relaxation time T_2 (i.e., the rate of coherentization must be greater than that with which coherence is destroyed).

Thus, two conditions must be satisfied for the self-excitation of the electromagnetic emission in the chemical reaction to take place: the negative polarization of the nuclei, to allow energy to be stored in the Zeeman reservoir, and the coherence of the nuclear spin system, to rescue Zeeman energy from thermal relaxation wasting and to stimulate its radiative emission. These two conditions are closely linked: if Zeeman energy is small, the EMF induced by the precession of **M** and its coherence will both be weak, and they will be unable to compete with the transverse nuclear relaxation processes that destroy the coherence. Therefore, to excite generation it is necessary to achieve a well-defined critical level of the nuclear polarization (the critical magnitude of vector **M**) corresponding to the amount of stored Zeeman energy called the generation threshold [8].

4.3. Experimental Observations of Radio-Frequency Emission

Chemically induced RF emission was first observed in the photochemical reaction between porphyrin and quinone [9], and somewhat later in the thermal decomposition of benzoyl peroxide [10].

The scheme of the reaction between porphyrin and *p*-benzoquinone is

$$\mathrm{Por} \xrightarrow{h\nu} \mathrm{Por}^*$$

$$\mathrm{Por}^* + \mathrm{Q} \xrightarrow{\mathrm{H}^+} \mathrm{Por}^{+\cdot} + \dot{\mathrm{Q}}\mathrm{H}$$

$$\mathrm{Por}^{+\cdot} + \dot{\mathrm{Q}}\mathrm{H} \longrightarrow \mathrm{Por} + \mathrm{Q} + \mathrm{H}^+$$

where Por is a porphyrin molecule (octamethyltetraphenyl porphyrin), Q denotes a quinone molecule, $\dot{\mathrm{Q}}\mathrm{H}$ is the semiquinone radical, and $\mathrm{Por}^{+\cdot}$ is the porphyrin radical cation. The electron transfer reaction produces ion radicals and their further disproportionation regenerates quinone molecules with negatively polarized protons.

The reaction was carried out in a glass tube placed in a circuit coil, and initiated by light in the porphyrin absorption band ($\lambda = 546$ nm). The circuit was tuned to the Larmor precession frequency of the proton magnetization vector of the quinone molecule $f_L = 100$ MHz ($H = 24$ kG), corresponding to the emission wavelength of 3 m. A block diagram of the detection circuit is shown in Fig. 4.5. To lower the generation threshold the Q-factor of the tuned circuit was increased by additional feedback from the high-frequency amplifier. After preamplification of the signal from the chemically induced RF field a low-frequency signal was obtained at the difference frequency $f_\mathrm{d} = f_\mathrm{L} - f_\mathrm{g}$, where f_g is the frequency of the reference generator, and this final signal was amplified by the low-frequency amplifier and displayed on a recorder or oscilloscope. The self-excitation after the high-frequency amplification was additionally monitored at the high frequency f_L by a high-frequency oscilloscope.

A record of the chemically induced signal is shown in Fig. 4.6. Generation begins 5–10 s after the photolysis starts (i.e., when the accumulated Zeeman energy reaches the generation threshold). After this short initial transient a

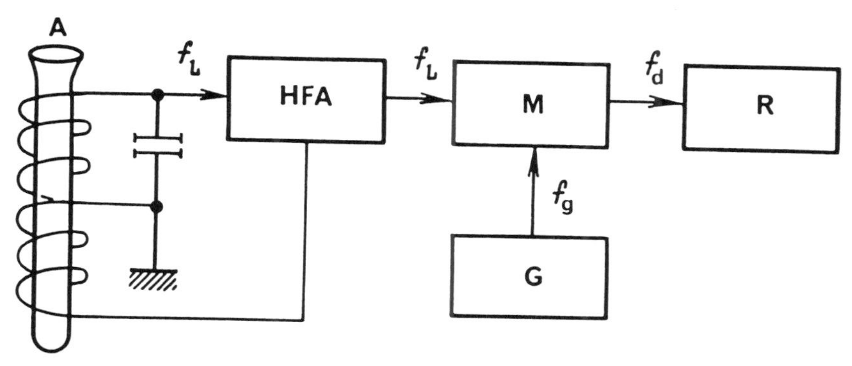

Figure 4.5 Block diagram of the mw circuit used to detect a chemically induced mw field. A, sample with polarized product molecules; HFA, high-frequency amplifier; M, frequency mixer; G, generator; R, recorder.

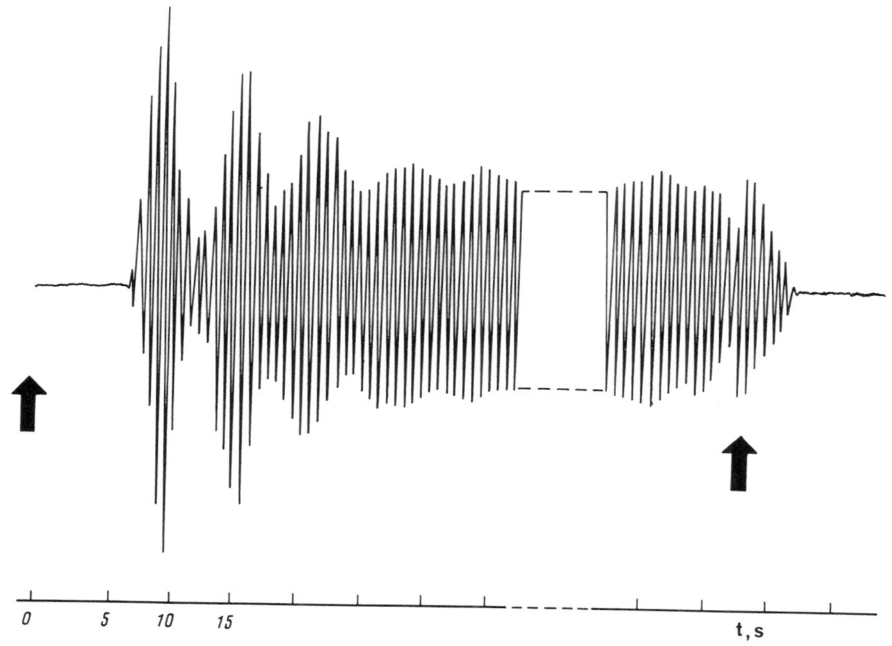

Figure 4.6 Chemically induced RF emission from quinone molecules, the photochemical reaction product. The signal is recorded at the frequency $f_L = 100\,\text{MHz}$ ($H = 24\,\text{kG}$). The arrows indicate the moments of switching on/off the photochemical pumping [9].

steady-state regime is established and is maintained for a long time, because the reaction is almost ideally reversible and therefore ensures continuous pumping of the nuclear spin reservoir. When the light is switched off the reaction stops. The generation process can be repeated many times by switching the photochemical pumping on and off.

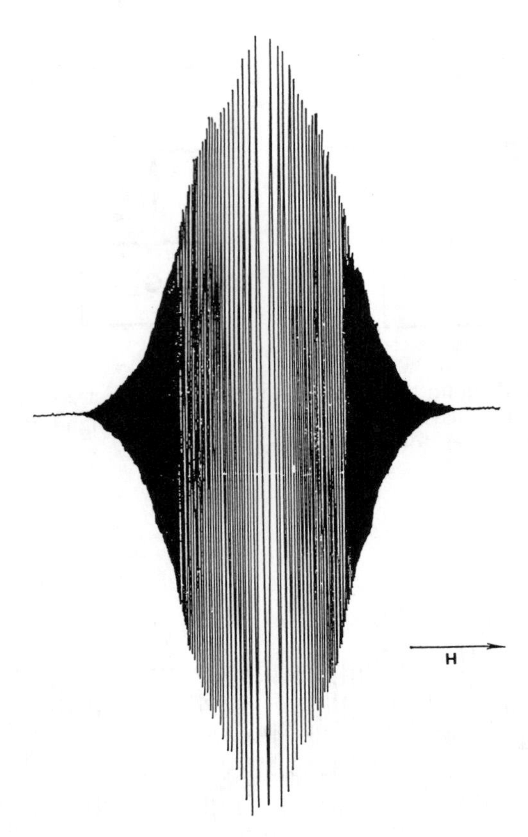

Figure 4.7 Chemically induced RF emission from quinone molecules recorded with periodic variations of the field.

Figure 4.7 shows a record of the RF generation obtained under the same conditions but with a slowly varying magnetic field (in the range of 10^{-3} G) with the recorder carriage connected to the field scanner. The field strength variation changes the generation frequency f_L, and hence the detected difference frequency f_d. Thus, $f_d = 0$ at the center of each bundle of signals, one of which is shown in Fig. 4.7. Here the Larmor precession frequency coincides with that of the reference generator f_g, and f_d grows up toward the edges of the bundle. This behavior of f_d evidences unequivocally that the signal is due to chemically induced radiowave emission.

In another series of experiments the RF field was detected in a crossed coil system (Fig. 4.8). The auxiliary coils are connected to the output of the high-frequency amplifier, and the generation signal enters into the input of the same amplifier. As in the previous case, H is 24 kG and the precession frequency is 100 MHz. The generation signals detected in this way are shown

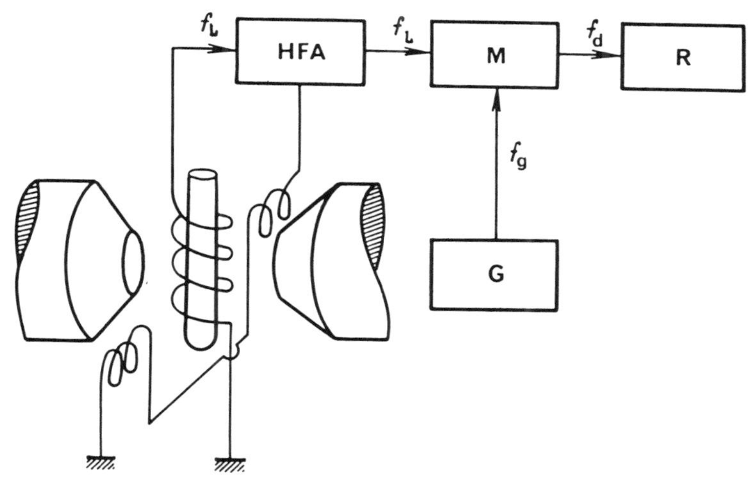

Figure 4.8 Block diagram of the crossed coils setup used to detect chemically induced RF emission: HFA, high-frequency amplifier; M, mixer; G, standard generator; R, recorder.

in Figs. 4.9 and 4.10. The former were recorded by amplitude detection with the field H kept constant, while the latter, shown in Fig. 4.10, were recorded with a slowly varying field H in each bundle of lines (similar to Fig. 4.7). The generation consists of regularly spaced short pulses (or bundles of pulses). In this regime the *reaction system behaves like a radio-frequency chemical pulsar.*

The amplitude of chemically induced RF field has been estimated in experiments on the self-excitation of the precession of the total nuclear

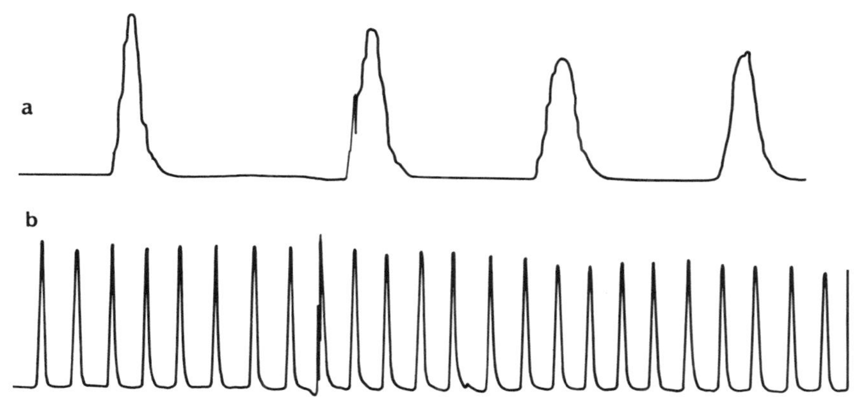

Figure 4.9 Chemically induced RF emission from quinone molecules in pulsar regime (H = const). The duration of the record is 50 s (a) and 250 s (b).

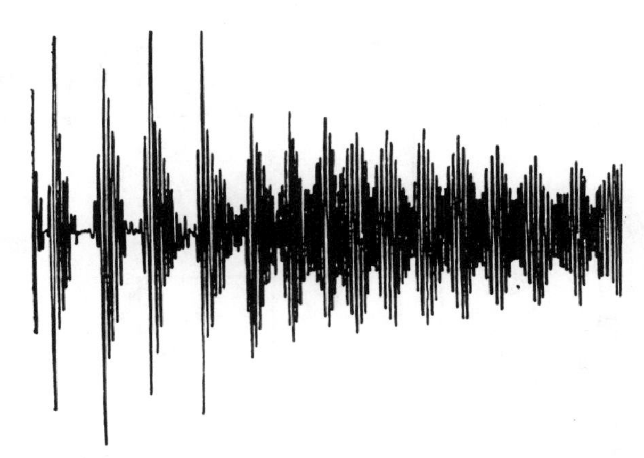

Figure 4.10 Chemically induced radio emission from quinone molecules with slow changing of magnetic field.

magnetization vector **M** in a rotating coordinate system, involving a nonresonant magnetic field H_1 rotating in a plane perpendicular to the constant field H. The RF field was generated by the negatively polarized protons of the benzene molecules produced by the thermal decomposition of benzoyl peroxide in cyclohexanone (120°C, peroxide concentration 0.5 mol). Under these conditions the negative polarization and RF generation are retained for 3–4 min. From the nuclear polarization kinetics (similar to that shown in Fig. 4.2) the rate constant κ ($2 \times 10^{-2}\,s^{-1}$) of the reaction

$$(PhCO_2)_2 \xrightarrow{\kappa} C_6H_6$$

and the polarization coefficient Z (-400) were determined. In these experiments the amplitude of the EMF produced in the coil was found to be $\approx 2\,\mu V$. It corresponds to the amplitude of the magnetic component of the EMF generated by the reaction of $\approx 1.6 \times 10^{-7}$ G. It allows us to estimate the emission temperature as $T_{em} \approx 10^{28}$–10^{29} K (i.e., the energy of the nonequilibrium emission from the reaction is many orders of magnitude higher than that of the equilibrium one). In comparison we note that the energy of the microwave radiation from one of the best known space emitters, the hydroxyl maser, is only $\approx 10^{13}$ K (i.e., the chemical RF generator is much brighter than its space analog). However, the absolute efficiency of the chemical generator is negligibly small, because the heat capacity of the spin systems is very low and only a very small fraction of the chemical reaction energy can be accommodated in the Zeeman reservoir.

4.4. Dynamics of Radio-Frequency Generation

The theoretical analysis of the chemical RF generator [11] is based on the standard Bloch equations, which describe the behavior of the nuclear magnetization **M**:

$$dM_x/dt = \omega M_y - \gamma_n M_z H_y - 2\pi\Delta f M_x \tag{4.13}$$
$$dM_y/dt = -\omega M_x + \gamma_n M_z H_x - 2\pi\Delta f M_y$$
$$dM_z/dt = \gamma_n(M_x H_y - M_y H_x) - \beta(M_z - M_0) + Z(dM_0/dt)$$

where M_x, M_y, M_z are the components of the magnetization vector, M_0 is the equilibrium magnetization, H_x and H_y are the components of the oscillating magnetic field generated by the chemical reaction, $\beta = T_{1n}^{-1}$, and Δf is the halfwidth of the NMR line. Equations (4.13) differ from the common Bloch equations only by the term $Z(dM_0/dt)$, which determines the rate of chemical pumping.

The self-excitation threshold is known from general radiophysics to be conditioned by

$$|M_z| > \Delta f/\gamma_n \eta q \tag{4.14}$$

where η is the filling coefficient of the coil and q is the Q-factor of the circuit. If the inequality (4.14) is satisfied, the self-excitation of the precession of the vector **M** occurs (Fig. 4.3). Under these conditions a transverse component of the magnetization $M_\perp = (M_x^2 + M_y^2)^{1/2}$ arises and increases exponentially. Therefore, the high-frequency EMF is induced in the coil with the amplitude given by

$$U = (4\pi/c)qNS_a\eta\omega_L M_\perp \tag{4.15}$$

where N is the number of turns in the coil, S_a is its cross section, and c is the speed of light.

For unimolecular reactions of the type of A → B, with the rate constant κ and the polarization coefficient Z the self-excitation condition becomes

$$\left[\left(\frac{\beta - Z\kappa}{\beta}\right)^{\beta/(\beta-\kappa)}(1 - Z)^{\kappa/(\kappa-\beta)}\right] > 1 + \frac{\Delta f}{\gamma_n \eta q M_0} \tag{4.16}$$

where some quantities (Z, κ, M_0, and β) are "chemical" parameters, and the others (Δf, η, and q) refer to radiophysics.

The generation dynamics obtained by solution of Eqs. (4.13) is illustrated in Fig. 4.11. In the case of pulsed chemical pumping (the fast reaction) the generation consists of a single pulse (Fig. 4.11a). The magnetization M_z quickly reaches the generation threshold, and then the transverse component $M_\perp$ appears. The relationship between M_z and $M_\perp$, the phase diagram, looks like a single loop (see inset in Fig. 4.11a) (i.e., the chemical pumping pulse is accompanied by an RF generation pulse).

With strong and continuous chemical pumping the RF emission takes the form of a regular series of pulses (Fig. 4.11b). Between the pulses the chemical reaction continues to create new magnetization, which reaches the generation threshold. The phase diagram consists of many superimposed loops. This regime was demonstrated experimentally (see Fig. 4.9, the *chemical pulsar*). If

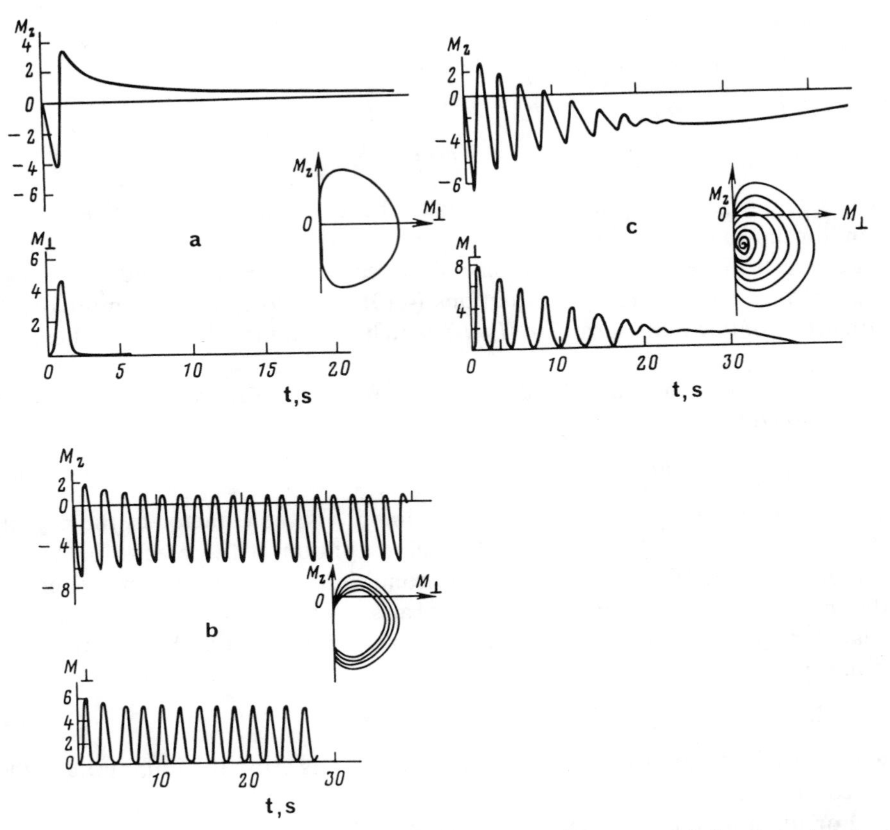

Figure 4.11 RF generation dynamics of the chemical raser: (a) for fast chemical pumping, (b) for continuous pumping (reversible photochemical reaction), (c) for pumping by slow unimolecular reaction. In each case the phase diagram of the nuclear magnetization in $M_{\perp}$, M_z coordinates is shown in the inset. The emission intensity is proportional to $M_{\perp}^2$.

the rate of chemical pumping is lowered the pulse regime may change into continuous steady-state generation, as shown in Fig. 4.6. In the case of damped chemical pumping, one can observe a smooth change in the regimes (Fig. 4.11c): the pulsed generation is replaced by continuous generation, which, in turn, decays as the chemical reaction attenuates. Depending on the kinetics of the chemical pumping, different generation regimes can be implemented; several of them have been demonstrated experimentally (see Section 4.3).

4.5. Chemical Raser

Chemical reaction as the radiowave emitter is a quantum generator, molecular broadcasting station, the *chemical raser*, an analog of laser and maser.

To transform a chemically reacting system into the chemical raser, it is necessary to satisfy two general requirements.

1. The energy stored in the nuclear spin system of the reaction products should exceed the generation threshold.
2. The coherent motion of negatively polarized nuclear spins is needed to ensure that this energy excess would not be wasted into the heat but be emitted.

The coherence is supported by the interaction between the magnetization and EMF generated in the coil circuit. However, the self-excitation of the RF generation is possible in principle even in the absence of any special resonance elements (i.e., in the free space). The conditions for this type of excitation were investigated [12], and the generation threshold was found to be defined by the equation

$$|M_z|V \geqslant 3c^3/(2\gamma_n T_2 \omega_L^3) \tag{4.17}$$

where V is the emitter volume. For example, the minimum volume of the free space emitter at a 1 mol concentration of the emitting molecules, the nuclear polarization Z_{lim}, $T_2 = 10\,\text{s}$, and $f_L = 100\,\text{MHz}$ ($H = 24\,\text{kG}$) equals $100\,\text{cm}^3$. Under the same conditions, but in the earth field $H = 0.3\,\text{G}$ $V_{\text{min}} \approx 10^{15}\,\text{cm}^3 \approx 1\,\text{km}^3$. These estimates show that the RF generation of objects such as man or other animals and the telepathy via chemical or biochemical pumping of nuclear spin systems of the living organisms are hardly possible.

The radio emission can be in principle produced by a system of negatively polarized electrons in radicals. The special features and properties of this chemically induced emission of electron Zeeman reservoir have been investigated [12].

References

1. Buchachenko, A. L. *Chemically Induced Polarization of Electrons and Nuclei*. Science, Moscow, 1974.
2. Buchachenko, A. L. *Russ. Chem. Rev.* **1971**, *40*, 1729.
3. Bubnov, N., Medvedev, V., Polyakova, K., Okhlobystin, O. International Symposium on CIDNP and CIDEP, Tallinn, 1972.
4. Levin, Ya., Ilyasov, A., Pobedimskii, D. *Proc. Sov. Acad. Sci. Chem. Ser.* **1970**, 1680.
5. Grishin, Yu., Dushkin, A., Sagdeev, R. *High Energy Chem.* **1978**, *12*, 278.
6. Grishin, Yu., Dushkin, A., Sagdeev, R. *Russ. J. Phys. Chem.* **1979**, *53*, 278.
7. Abraham, A. *The Principles of Nuclear Magnetism.* Clarendon Press, Oxford, 1961.
8. Buchachenko, A. L., Berdinskii, V. L. *Russ. Chem. Rev.* **1983**, *52*, 3.

9. Zhuravlev, A. G., Berdinskii, V. L., Buchachenko, A. L. *JETP Lett.* **1978**, *28*, 140.

10. Berdinskii, V. L., Buchachenko, A. L. *Russ. J. Phys. Chem.* **1981**, *55*, 1921.

11. Berdinskii, V. L., Buchachenko, A. L., Pershin, A. D. *Theor. Exp. Chem.* **1976**, *12*, 666.

12. Berdinskii, V. L., Buchachenko, A. L. *Russ. J. Phys. Chem.* **1980**, *54*, 386.

CHAPTER

5

Reaction Yield Detected Magnetic Resonance—RYDMR

5.1. Principle of Chemical Reception of Electromagnetic Waves

5.1.1. Absorption of Electromagnetic Waves

The influence of electromagnetic waves on the course of chemical reactions is assumed to be caused by an interaction of the magnetic field B_1 of the electromagnetic wave with elementary magnetic moments of particles that take part in the reactions. The interaction may in principle change the state of the particles, thus influencing the probability of reaction between them. In this respect the influence of the electromagnetic power absorbed is similar to chemical effects of the light as studied by photochemistry. Electronically and vibrationally excited species are produced under the action of the light, thus changing their reaction ability due to excitation. As previously stressed, the course of the reaction depends not only on the energy of active particles but also on their spin state. One cannot rely on any increase of energy of active particles as a result of the radiowave power absorption as an energy of a single quantum is much lower than thermal motion energy kT: at temperature $T = 300\,\mathrm{K}$ $kT = 6 \times 10^{-21}\,\mathrm{J}$, whereas the energy of a quantum for radiowaves with wavelength λ situated between 1 cm and 100 m corresponds to energies from 2×10^{-23} to $2 \times 10^{-27}\,\mathrm{J}$. Thus our primary attention will be on a possible change of spin states of active particles under the action of radiowaves, and the chemical sequences of that change.

5.1.2. Equilibrium Population of Energy Levels

To absorb the energy of electromagnetic waves the irradiated system must have a number of energy levels, and the population density of these levels must be nonequal. Otherwise transitions will occur with the same rate in both directions (both with absorption and with emission of the energy) and the state of the system will remain unchanged. The ratio of an energy absorbed to an energy that is incident on a two-level system and induces transitions between levels is estimated as $\delta W/W = p$, where $p = (n_1 - n_2)/(n_1 + n_2)$ is the polarization ratio, expressed through population densities n_1 and n_2 of two levels (see Fig. 5.1). The value of p depends on the rate of spontaneous transitions between levels, $k_\downarrow$ and $k_\uparrow$ ($k_\uparrow = k_\downarrow \exp(-\Delta E/kT)$, ΔE is an energy interval between levels). The rate of transitions induced by an external electromagnetic field B_1, $k^* \sim B_1^2$. It is possible to show that

$$p = (1/2)(k_\downarrow \Delta E/kT)/(k_\downarrow + k^*) \quad (5.1)$$

To make an estimation let us take $\Delta E = h\nu = 2 \times 10^{-24}$ J (electromagnetic wave with wavelength $\lambda = 10$ cm). Then at $T = 300$ K we shall get

$$\delta W/W = p \leqslant (1/2)h\nu/kT = 2.5 \times 10^{-3}$$

Thus at room temperature the energy of the electromagnetic wave (of radiowaves) may be absorbed by the two-level systems, although very inefficiently. The hope of changing the state of active particles in such conditions is very weak.

The value of p may be increased by lowering the temperature. If the system will be irradiated at such temperature that $T < \Delta E/k$ it is possible that nearly all the transitions induced will absorb the energy. At $\Delta E = h\nu = 2 \times 10^{-24}$ J the last condition is fulfilled at $T = 0.1$ K. Cooling of the system down to liquid helium temperature (4 K) often leads to high enough polarization.

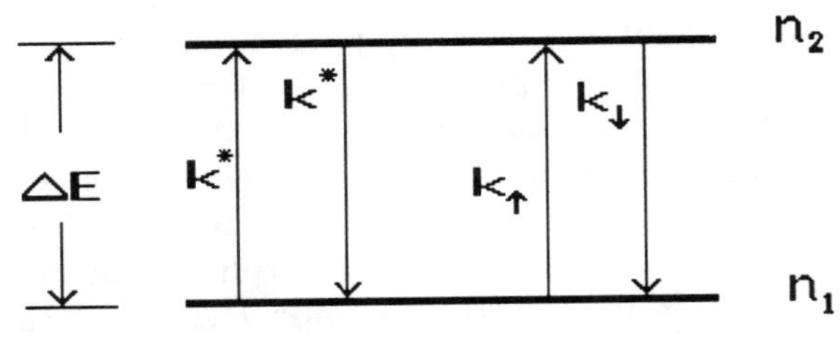

Figure 5.1 Transitions in a two-level system. $k_\downarrow$ and $k_\uparrow$ are rate constants of spontaneous transitions, k^* ($\sim B_1^2$) is the rate constant of transitions induced by electromagnetic waves with a.c. magnetic field B_1; n_1 and n_2 are population densities of levels; ΔE is an energy gap between levels.

5.1.3. Creation of Nonequilibrium Population

A difference in the population densities of the neighbor energy levels may be produced dynamically in a nonequilibrium process of creation of excited states from an independent source. Such a process is shown in Fig. 5.2 with an example of the creation of excited triplet states of molecules. Steady-state populations of magnetic sublevels T_x, T_y, and T_z are determined by the corresponding rates of population g_x, g_y, and g_z and rates of decay k_x, k_y, and k_z. Relaxation processes that are characterized by the time constants T_1 T_2, tend to make an equilibrium Bolzmann distribution of population densities. However at $T_{1,2} \gg 1/k_{x,y,z}$ (*) the densities will be equal to $n_x = g_x/k_x$, $n_y = g_y/k_y$, and $n_z = g_z/k_z$. At low temperature the condition (*) is easier to fulfill as relaxation times increase when T becomes lower.

An absorption of electromagnetic power as a result of transitions between magnetic sublevels of triplet molecules at liquid helium temperature is the basis of well-known magnetic resonant methods used for investigation of these molecules. Using electron paramagnetic resonance (EPR) a microwave power absorption rate caused by the transitions is measured. The sample under investigation stays in the external permanent magnetic field B_0, which tunes the energy difference between levels ΔE in resonance with the quantum energy: $\Delta E = g\beta B_0$, where g is the g-factor of the paramagnetic particle and β is the Bohr magneton.

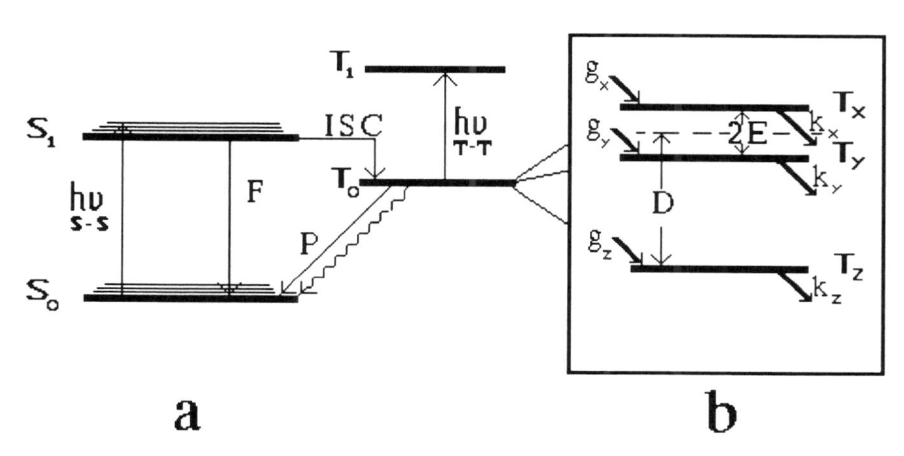

Figure 5.2 Production of the triplet excited state T of the molecule from the singlet excited state S by intersystem crossing (a). On the right side (b) a fine structure of the triplet state is shown, and rates of population and decay are indicated. D and E are parameters of the fine splitting.

5.1.4. ODMR Techniques

Optically detected magnetic resonance (ODMR) is based on the monitoring of a change of optical parameters of the sample induced by the absorption of radiowave power [1]. Magnetic sublevels usually have different rate constants for spontaneous transitions from excited to ground state. Absorption of electromagnetic power leads to a change in their population densities. One ODMR technique uses a change in the intensity of phosphorescence that accompanies the transition from triplet excited to singlet ground state [phosphorescence detected magnetic resonance (PDMR)]. Another technique uses an absorption of the probing light, which is able to produce triplet–triplet transitions. The absorption is proportional to the difference in the population densities of the sublevels [absorption detected magnetic resonance (ADMR)] (see Fig. 5.2a).

ODMR spectroscopy primarily permits us to learn about splitting of magnetic sublevels in the triplet state of the molecules in zero external magnetic field, about an anisotropy of that splitting. It provides knowledge on the parameters of fine interaction tensors, **D** and **E**. It also gives some kinetic parameters of triplet excitations, and rate constants for radiative and non-radiative decay, relaxation times T_1 and T_2, and their dependence on the surrounding of triplet molecules. Only solid state samples may be used in ODMR. Fast rotation of molecules averages to zero magnetic dipole interaction, and ODMR ceases to work. If triplet molecules take part in a reaction then the rate of this reaction may change as a result of a change in the population density on some magnetic sublevel. There is, however, only a few examples of such reactions due to the necessity to work at low temperature (see Leung and El-Sayed [2], who observed photodecomposition of pyrimidene at 1.6 K from the triplet excited state, the rate of which depended on the magnetic resonant transitions in the triplet state).

5.1.5. Bimolecular Reactions in Condensed Phase

These reactions appeared to be more promising for "chemical detection" of radiowaves. When a particle that has a magnetic moment interacts with the magnetic field of electromagnetic wave, and absorbs the energy, a change of the spin state of the particle usually occurs. Projection of the magnetic moment vector on the chosen direction is a characteristic of the state and can vary, whereas the value of the magnetic moment remains unchanged. For a change of the spin state of the particle to reveal itself in the chemical activity, the counterpart of the particle must also have a magnetic moment. In such a case a pair of active particles having magnetic and mechanical momenta will take part in the reaction. And the reaction itself will obey the momentum conservation law: the reaction rate will be spin dependent.

It must be stressed that this law is valid for isolated systems only. Considering reactions in the condensed phase it is not easy to imagine their elementary steps as isolated from the surroundings: collision frequencies of the particles are of order of $10^{13}\,s^{-1}$. However, it is worth mentioning that only spin motion is important in connection with the influence of the magnetic moment. Disturbance of the spin motion due to interaction with neighboring molecules is characterized by rather long times: by time constants of spin–lattice relaxation T_1 and spin–spin relaxation T_2. Their values even at room temperature are of order 10^{-6} to $10^{-8}\,s$. During the time interval that is shorter than these values the reactions in the condensed phase in respect to their spin degrees of freedom may be considered as isolated from the environment.

Reactions in which pairs of paramagnetic particles take part are as a rule exothermic and fast. It means that they produce products in one collision if the law of momentum conservation is fulfilled. Otherwise the reactions do not go.

Within the past few decades great success in developing methods of investigation of the mechanism of bimolecular reactions in the condensed phase was connected with the discovery of the possibility of modulating their rate by external magnetic fields. There are a few review papers devoted to this topic (see, e.g., the most complete and recent review of Steiner and Ulrich [3], and also [4–7]). This topic was also considered in Chapter 3. The key point of the magnetic field effect was shown to consist in the influence of the external magnetic field on the character of evolution of the total spin of the pair of paramagnetic species, which are in the position when they are ready to react if the spin conservation law will be fulfilled. Spin evolution switches the possible ways of the reaction, on each of them the spin being conserved, and thus changes the yield of products of the reaction. The depth of the change depends on the initial spin state of the pair, on its lifetime in "ready-to-react" state (or lifetime in the cage), and on the rate of spin evolution.

5.1.6. Spin Evolution of the Pair of Paramagnetic Particles

The process of spin evolution in a pair is governed by interaction affecting the degree of freedom of the spin. The main interactions are the following. First, mention should be made of the Zeeman interaction with an external magnetic field in the case of different magnetic moments of the pair members. This is the so-called Δg-mechanism. An *isotropic hyperfine interaction (HFI)* of the unpaired electron with magnetic nuclei constitutes another mechanism. For triplet species an effective mechanism for mixing spin states is the dipole–dipole interaction of electron spins. Among a coherent spin conversion there may be an interaction induced by randomly modulated coupling that has the character of spin relaxation; this is referred as the *relaxation* mechanism. The rate of relaxation generally depends on the strength of the external magnetic field.

5.1.7. Change of the Spin State of the Pair by Electromagnetic Waves

For the action of electromagnetic waves on the reaction the basic fact is that the spin state of a pair of species can be changed not only by a constant external magnetic field but also as a result of radiowave magnetic field-induced transitions between Zeeman sublevels of one constituent of the pair. In that case absorption of electromagnetic radiation energy directed onto a system undergoing a pair reaction will affect the reaction yield; so the spectrum of the effect on the reaction yield will correspond to a spectrum of radiowave absorption. Since the reactivity of paired active particles changes with respect to one another because of resonant absorption, it is possible to record the spectrum of intermediate short-lived states. It is this possibility that underlies the method of reaction yield detected magnetic resonance (RYDMR). The recorded parameter related to the reaction yield may be any readily measurable quantity. Most commonly it is the luminescence intensity if the reaction products are electron excited states. Even when using optical recording techniques, the RYDMR method differs essentially from the ODMR methods mentioned above. To account for magnetic resonant phenomena in ODMR one can simply consider an isolated paramagnetic species (generally a triplet) with unequally populated magnetic sublevels that undergoes population redistribution caused by electromagnetic absorption. As for the RYDMR method, a requisite condition is the reaction in a pair of paramagnetic species, at least one of which is mobile, that changes their reactivities because of a changed spin state.

5.2. Reaction Yield Detected Magnetic Resonance—RYDMR

5.2.1. Generation and Decay of Pairs

The reaction step that is spin dependent appears when two paramagnetic particles meet each other. Such particles may be free radicals or ion radicals, molecules or excitons in the triplet state in crystals, electrons and holes in semiconductors, or paramagnetic defects in polymers or other solids. Figure 5.3 displays a scheme of such a step. A pair of paramagnetic particles R_1 and R_2 is produced as a result of the transformation of precursors and has a spin state that corresponds to that of the total spin of precursors. Being in the condensed phase particles R_1 and R_2 spend some time in the vicinity of each other during which they have a chance to approach each other at distances where a chemical reaction can start. If in the course of the reaction some products may be produced that have the same spin as that of the pair, then the reaction proceeds. Otherwise the particles of the pair do not interact and go apart. During the time the particles stay in the vicinity of each other (in the reaction

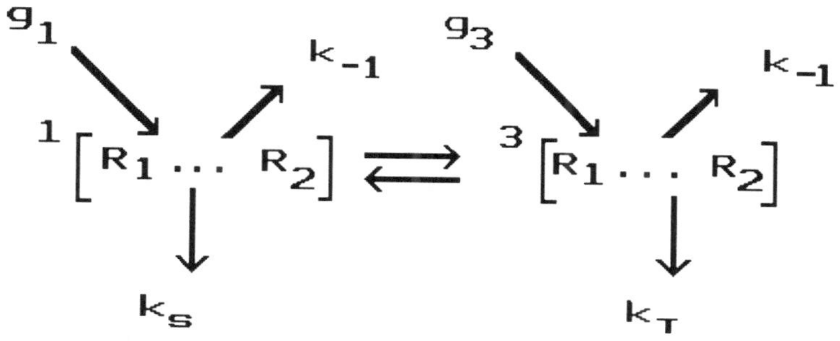

Figure 5.3 A scheme of production and decay of a pair of paramagnetic particles R_1 and R_2 each having spin $S = 1/2\hbar$. g_1 and g_3 are rates of production of pairs from precursors. Rate constants k_S and k_T refer to reactions proceeding at the interaction of particles R_1 and R_2 in singlet and triplet channels. k_{-1} is the rate constant of the dissociation of pairs.

cage) the total electron spin of the pair changes (evolutes) due to different magnetic interactions. A pair that survived after an attempt to enter into reaction has a chance to react again in a new spin state.

Investigation of the behavior of paramagnetic species is usually carried out in the external permanent magnetic field. When a number of conditions (listed below) are fulfilled, the magnetic field influences evolution of the total spin substantially. As a result of such an influence the yield of the products changes and magnetic field spin effect takes place. At present many processes where paramagnetic species are involved were found to be affected by the magnetic field. The same processes are sensitive to an a.c. magnetic field of electromagnetic waves. A typical reaction in the pair of ion radicals produced during the interaction of excited molecule with an electron acceptor may be considered as an example.

This reaction takes place in systems containing light-absorbing molecules on production of electron excited species. For example, on excitation of molecular crystals consisting of molecules $^1D_0$ and having impurity molecules $^1A_0$, the reaction is accompanied by an electron transfer from $^1D_0$ to $^1A_0$ by the appearance of charge transfer pair states $D^+ \cdots A^-$, which may either recombine or dissociate into free charges:

$$^1D_0 + {}^1A_0 \rightarrow [D^+ + A^-] \rightarrow D^+ + A^- \tag{5.2}$$

Hereafter we will use the following notations for the electron states, ^pA_q where A (or any other letter) stands for molecules or excitons, p represents the multiplicity of the state, and q is the number of the excited state assuming that for the ground state $q = 0$. Reaction (5.2) is actually a general scheme of an oxidation–reduction reaction. In liquid polar solvents containing solute molecules $^1D_0$ and $^1A_0$ such a reaction will produce exciplexes that would then dissociate into ionic pairs [which are equivalent to the $(D^+ + A^-)$ stage in Reaction (5.2)] and then into free ions. In molecular crystals species such as

$^1D_1$ and D^+ may be considered as movable excitons and holes whereas $^1A_0$ and A^- remain localized as impurity molecules and trapped electrons. According to Reaction (5.2) an initial state of the pair ($D^+ + A^-$) is a singlet.

5.2.2. Spin States of the Pair

Let us consider the behavior of spins of the system in the magnetic field (Fig. 5.4a). Vectors represent magnetic moments $\mathbf{M}_1$ and $\mathbf{M}_2$ of D^+ and A^-. A pure singlet state of the pair is pictured as two vectors directed in opposite sides, whereas triplet states are produced by different orientations of vectors giving a nonzero sum of the vectors.

It should be noted that the vector presentation of the spin states is based on the description of the spin state of an electron by functions α and β corresponding to the orientation of spin along the chosen direction and against it. For two electrons of the pair the spin states are described as combinations of α and β for each of the electrons. Function $\alpha_1\alpha_2$ corresponds to the T_{+1} state and function $\beta_1\beta_2$ corresponds to T_{-1} in total agreement with the vector scheme. However, the vector presentation of states S and T_0 is not completely correct as it assumes a certain direction of the spin or a certain radical (e.g., in Fig. 5.4 the radical with a magnetic moment M_1 has the direction α_1, and the radical with M_2 has β_2, the pair is described by the function $\alpha_1\beta_2$; but the description of the state as $\beta_1\alpha_2$ would look equally substantiated). A correct description of S and T_0 states consists in a superposition of the two polarized states:

$$|S\rangle = \frac{1}{\sqrt{2}}(\alpha_1\beta_2 - \beta_1\alpha_2),\ |T_0\rangle = \frac{1}{\sqrt{2}}(\alpha_1\beta_2 + \beta_1\alpha_2)$$

Such a superposition is not easy to express by a simple two vector scheme. However, a proper qualitative presentation of the main properties of spin states of the pair may be obtained from the two vector scheme. An energy diagram of the pair of paramagnetic species with each member of the pair having spin $s = 1/2\hbar$ in an external magnetic field is presented in Fig. 5.5.

There are a few mechanisms of spin evolution of the pair: *In an external permanent magnetic field* B_0:

1. Zeeman interaction $H = 1/2g\beta B_0$. It may differ for two particles D^+ and A^- due to a difference in the values of their magnetic moments $M_{1,2} = g_{1,2}se/(2mc)$. It causes transitions between states S and T_0 with frequency $\Delta\omega = 1/2\Delta g\beta B_0$ where $\Delta g = |g_1 - g_2|$, producing in the steady state mixed $(S\text{–}T)_0$ state of the pair.
2. Hyperfine interaction $H_{hfi} = \Sigma a_n m_n$(a_n is the isotropic hyperfine coupling constant of the nth nucleus, m_n is the z projection of the corresponding nuclear spin). A magnetic moment of the nuclei that belong to the ion radical D^+ produces a local magnetic field that disturbs the synchronized precession of spins and produces mixing of the S and T_0 states with a frequency $\Delta\omega = 1/2g\beta a_{eff}$ where a_{eff} is an effective hyperfine magnetic field.

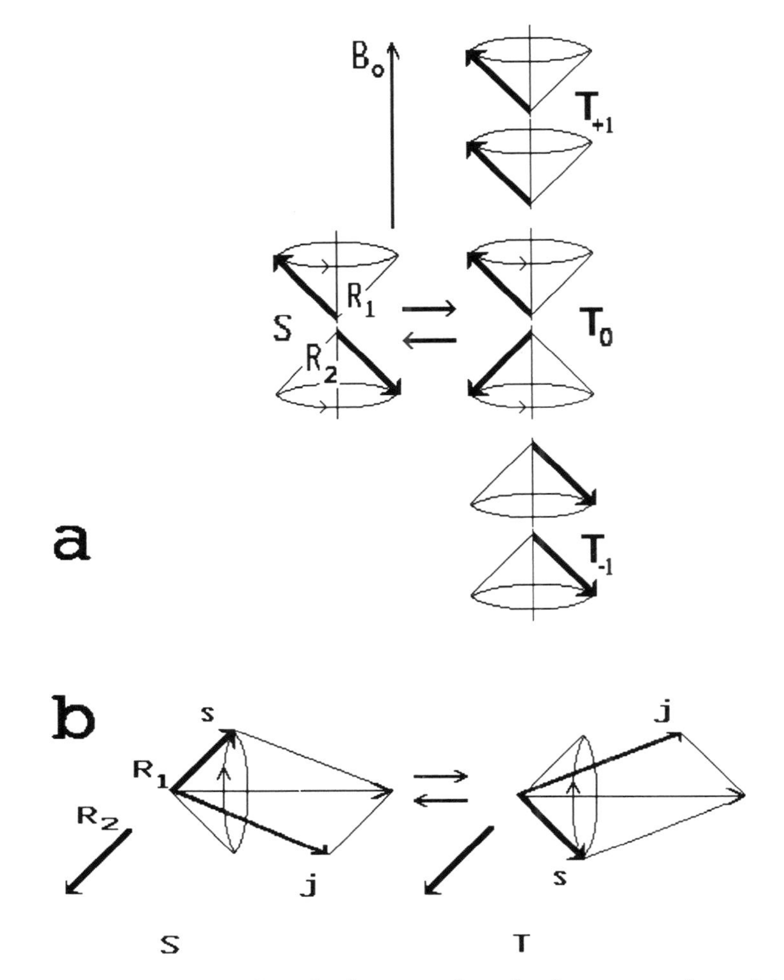

Figure 5.4 Vector presentation of spin states of a pair of paramagnetic particles with spin $S = 1/2\hbar$, and transitions between them. (a) Transitions between states S and T_0 in an external magnetic field are induced by a difference of precession frequencies of the magnetic moments around the direction of the field; (b) in the absence of the external magnetic field transitions between states are induced by hyperfine mechanism: transition between singlet and triplet states is shown to be caused by precession of electron spin S around the sum momentum vector; *j* is a nuclear spin.

3. Relaxation of spins caused by interaction with the surroundings of the pair may produce mixing of all the states during a time that is comparable or longer than spin–lattice T_1 or spin–spin T_2 relaxation times. Typical values of those are about 10^{-6} to 10^{-7} s at room temperature in a molecular media. Spin–lattice relaxation mixes T_0, T_{+1}, and T_{-1} states; spin–spin relaxation deals with mixing of S and T states. Values of T_1 and T_2 are magnetic field dependent.

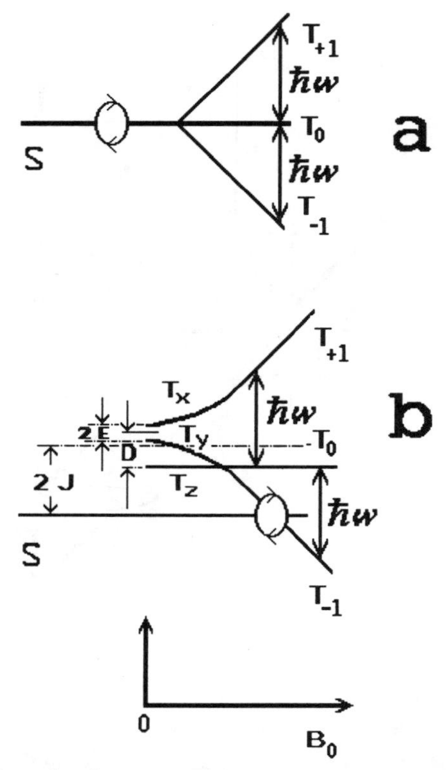

Figure 5.5 Energy of a pair of paramagnetic particles with spin $S = 1/2\hbar$ as a function of an external magnetic field. Two possible cases are shown. (a) Energy in the absence of interaction between particles; (b) exchange and dipole–dipole interactions are present; an external magnetic field is directed along the line connecting the particles. Arrows marked by $\hbar\omega$ show possible resonant transitions between triplet sublevels of the pair. Circles show the regions where mixing of singlet and triplet states is possible.

In the absence of an external magnetic field:

1. Zeeman interaction disappears.
2. Hyperfine interaction is able to mix all the spin states of the pair as shown in Fig. 5.4b. Precession of the electron magnetic moment around the vector, which is the sum of electron and nuclear spin, permits interconversion of any spin state into any other; typical frequencies of mixing remain the same as in the presence of an external field.
3. Relaxation is able to mix all the spin states for long living pairs.

Population densities of the spin states shown in Fig. 5.5 for pairs that have lifetimes shorter than relaxation time are far from equilibrium. A system of kinetic equations that takes into account generation, decay of pairs, and transitions between states may be used for calculation of the population densities. Such an approach is valid for pairs with first-order decay. A more general approach taking into consideration the evolution of spin, spin–dependent reactions, and motion of particles in pairs is based on the solution of the stochastic Liouville equation. However, to obtain a qualitative impression on the situation it is enough to understand that in conditions of production of pairs in a strong magnetic field from singlet precursors and effective transitions between S and T_0 states the population density of mixed ST_0 state will be higher than that of T_{+1} and T_{-1} states (see Fig. 5.5a). Resonant transitions between levels T_0 and T_{+1}, T_{-1} induced by radiowaves are possible in such a nonequilibrium situation. The transitions connect the mixed ST_0 state with states T_{+1}, T_{-1} and thus increase the rate of transitions $S \rightarrow T$. These transitions quench the singlet state of the pair, and if one monitors the reaction products in the singlet channel one can expect a decrease in their yield.

Exchange interaction that appears at small distances between active particles shifts singlet in respect to triplet levels. Mixing of the states is possible now only in the region of level crossing at nonzero external magnetic field (see Fig. 5.5b). When one considers the evolution of the spin of the pair for interparticle distances that are higher than 5 to 10 Å an exchange interaction may usually be neglected.

Absorption of electromagnetic power that produces transitions $T_0 \rightarrow T_{+1}$, T_{-1} has a spectrum that corresponds to a sum of transitions between Zeeman levels of two radicals of the pair. By varying the induction of the external magnetic field B_0 that is superimposed on the sample one may obtain the spectrum of electron paramagnetic resonance monitoring the yield of any reaction products connected with reactions in pairs. Spectra obtained in such a way were suggested [8] to call reaction yield detected magnetic resonance (RYDMR) spectra. The intensity of fluorescence that appears in the singlet channel of the reaction (e.g., at recombination of ion radicals) is often convenient to measure. Although the possibility of action by radiowaves affecting the rate of reactions in the pair of paramagnetic species dates back to 1966 [9], the first theoretical evaluation of the conditions in which a microwave magnetic field might substantially affect the process involving short-lived pairs of paramagnetic species has shown that if the pair lifetime is less than 10^{-7} s the microwave magnetic field should have $B_1 > 10$ G [10]. However, in principle, sensitive detectors permit obtaining the absorption spectra of microwaves from very small relative changes of the reaction yield.

Transitions that are necessary for changing the spin state of the pair proceed as a result of interaction of the magnetic field B_1 of the electromagnetic wave with the magnetic moment of a pair. Figure 5.5 shows these transitions as arrows between sublevels T_0 and T_{+1}, T_{-1}. At the same time these transitions correspond to a change of a spin state of one particle of the pair. Therefore the spectrum measured by the reaction yield is a sum of EPR spectra of particles of the pair. In particular, it contains hyperfine components for both particles of the pair. But the spectrum also has some features connected with properties of pairs: with a short lifetime of the pair before reaction or dissociation occur, and with the possible existence of exchange and dipole–dipole interaction between particles. Finite value of the lifetime determines the width of magnetic sublevels. A homogeneous width of the levels can be expressed by using rate constants k_S and k_T for reactions in the singlet and triplet channels, respectively, and k_{-1} as a dissociation rate constant:

$$\delta E \cong (k_S + k_{-1})\hbar \quad \text{or} \quad (k_T + k_{-1})\hbar \tag{5.3}$$

Exchange interaction gives a splitting $2J$ between singlet and triplet energy levels. If the value $2J$ greatly exceeds the width of the levels, then the mixing of singlet and triplet states becomes impossible. Small exchange interaction may reveal itself in splitting of the RYDMR signal and in peculiar effects of reversing of RYDMR spectra when the microwave power is increased. These effects will be described later. The exchange interaction decreases sharply with the distance r between particles:

$$J = J_0 e^{-r/a} \tag{5.4}$$

Typical values are $J_0 = 1$ eV, $a = 1$ Å. Evolution of spin of the pair for particles able to diffuse usually proceeds in the absence of exchange interaction. However, it switches on at a close enough approach of the particles, and it is responsible for the dependence of the reaction rate on the spin. For particles of the pair with fixed position in solid matrix a small exchange interaction may exist permanently during the evolution of spin.

Dipole–dipole interaction between particles may lead to splitting of RYDMR spectra (Fig. 5.5b). The interaction depends on the angle θ between the direction of an external field and a radius r connecting particles

$$E_{\text{d-d}} = (\mu_1\mu_2/r^3)(1 - 3\cos^2\theta) \tag{5.5}$$

In liquid solutions with fast rotational diffusion of pairs this interaction averages to zero (Fig. 5.5a and b).

5.2.3. Evaluation of the Magnitude of the RYDMR Signal

The relative variation in the yield γ of the reaction in pairs due to resonant transitions is determined by the product of probability p of the transition under the action of the microwave magnetic field B_1 and of the pair life time τ:

$$\Delta\gamma/\gamma = \alpha p \tau$$

For radicals with spin $s = 1/2\hbar$

$$p = 1/2\left(\frac{g\beta}{\hbar}\right)^2 \frac{B_1^2\tau}{1 + (g\beta/\hbar)(B_0 - B_{0\text{res}})^2} \tag{5.6}$$

The coefficient α (<1) provides relative variation in the yield if the multiplicity of the pair is completely changed. Under resonant conditions ($B_0 = B_{0\mathrm{res}}$)

$$\Delta\gamma/\gamma = \alpha\omega_1^2\tau^2 \tag{5.7}$$

where $\omega_1 = g\beta B_1/\hbar$. One can readily find that for pairs with a life time $\tau = 5 \times 10^{-9}$ s a $\Delta\gamma/\gamma$ of about 1% is achieved at $B_1 \cong 1$ G (at $\alpha = 1$). A minimum detected value of $\Delta\gamma/\gamma$ depends on the parameter of the reaction measured. For example, in measurement of luminescence intensity a minimum value of $\Delta\gamma/\gamma$ is specified only by the number of photons N that entered a detector over a period of detection. They produce the signal and a shot noise and

$$(\Delta\gamma/\gamma)_{\min} = N^{-1/2} \tag{5.8}$$

So for a recording device with a response time $t_{\mathrm{resp}} = 5$ s, and an intensity of luminescence that reaches the device equal to $I_{\mathrm{lum}} = 2 \times 10^9$ photons/s $N = I_{\mathrm{lum}} t_{\mathrm{resp}} = 10^{10}$ and $(\Delta\gamma/\gamma)_{\min} = 10^{-5}$.

5.2.4. Vector Model for Transitions in Pairs Induced by a Microwave Magnetic Field

For absorption of the energy from electromagnetic waves a certain polarization of the waves is necessary: vector $\mathbf{B}_1$ of the wave must lie in the plane that is normal to the direction of the external permanent magnetic field $\mathbf{B}_0$.

A vector model for transitions under the action of an a.c. magnetic field is shown in Fig. 5.6. Magnetic moment M_1 of the particle with a spin $s = 1/2\hbar$ ($M_1 = 1/2g_1\beta$) precesses with a Larmor frequency $\omega = 1/2\gamma B_0$, around the direction of the external permanent magnetic field $\mathbf{B}_0$, which is directed along the z-axis; here $\gamma = g_1\beta/\hbar$ is the gyromagnetic ratio. The vector of an a.c. magnetic field is directed along the y-axis. Its value is $2B_1^{\sim} = 2B_1 \cos \omega t$. The motion of the vector $2B_1^{\sim}$ may be decomposed onto two components with circular polarization, namely onto two vectors $\mathbf{B}_1$, rotating in the xy plane around the z-axis with a frequency ω in a direction opposite to each other. One of the rotating vectors appears to be immobile in the new system of coordinate x', y', z, which is rotating together with vector $\mathbf{M}_1$. In the rotating frame the vector $\mathbf{M}_1$ can feel only the permanent now magnetic field of the radiowave $\mathbf{B}_1$, and it rotates around its direction with a frequency $\omega_1 = 1/2\gamma B_1$. This leads to periodic reorientation of the vector $\mathbf{M}_1$ in respect to the z-axis and hence to the external field $\mathbf{B}_0$. Magnetic moment $\mathbf{M}_2$ of the second particle of the pair also suffers reorientation only if its frequency of rotation around the vector $\mathbf{B}_0$ is equal to ω. Looking at Fig. 5.4 it is easy to see that these reorientations correspond to transitions $T_0 \rightleftharpoons T_{+1}$ or $T_0 \rightleftharpoons T_{-1}$ between sublevels of the pair.

If the precession frequencies of vectors $\mathbf{M}_1$ and $\mathbf{M}_2$ in the field B_0 differ, then a feature in the spin evolution appears. Let us consider that feature.

As was mentioned before the difference between precession frequencies of magnetic moments may be caused by nonequal g-factors of the particles [Δg

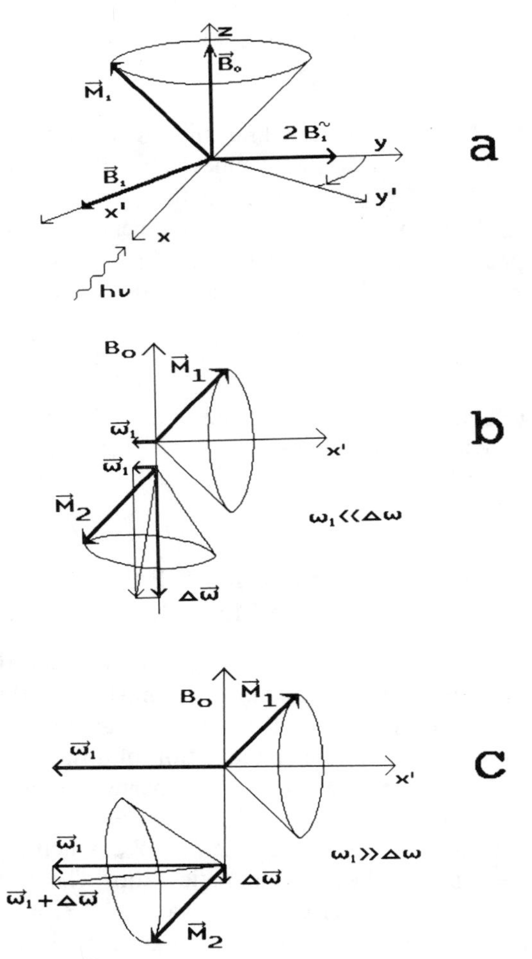

Figure 5.6 Vector model of the action of electromagnetic waves on the spin state of radicals with magnetic moments $\mathbf{M}_1$ and $\mathbf{M}_2$. (a) Precession of the vector $\mathbf{M}_1$ in a permanent magnetic field B_0. xyz is the laboratory system of coordinates; $x'y'z$ is the frame, rotating with resonant frequency $\omega = 1/2g\beta B_0/\hbar$. (b) Rotating frame; $\Delta\omega$ is the difference of precession frequencies of vectors $\mathbf{M}_2$ and $\mathbf{M}_1$; $\omega_1 = 1/2g_1\beta B_1/\hbar \ll \Delta\omega$. Precession of the vector $\mathbf{M}_2$ with a frequency $\Delta\omega$ mixes states S and T_0. Microwave magnetic field B_1 induces transitions $T_{-1} \rightleftharpoons T_0$ with a frequency ω_1. (c) Rotating frame; $\omega_1 \gg \Delta\omega$. Vectors $\mathbf{M}_1$ and $\mathbf{M}_2$ precess around the vector $\boldsymbol{\omega}_1$, mixing states S and T. Precession of vectors $\mathbf{M}_1$ and $\mathbf{M}_2$ around the direction of an external magnetic field $\mathbf{B}_0$ does not occur.

mechanism, $\Delta\omega = 1/2(g_1 - g_2)\beta B_0/\hbar]$ or/and by hyperfine interaction (HFI mechanism, $\Delta\omega = 1/2g\beta a/\hbar$, where a is a hyperfine interaction constant, in magnetic field units). In the frame x', y', z, rotating with a frequency ω, equal to the precession frequency of vector $\mathbf{M}_1$, this vector will be immobile, and vector $\mathbf{M}_2$ will precess around the z-axis with a frequency $\Delta\omega$. The frequency of mixing S and T_0 states (see Fig. 5.6b) will be the same. Switching on the radio-frequency field of a small amplitude B_1 when $\Delta\omega \gg \omega_1$ gives rise to $T_0 \rightleftharpoons T_{+1}, T_{-1}$ transitions, which can be seen as a normal RYDMR signal, say, as quenching of recombinational fluorescence from the singlet channel of the reaction in the pair. At a higher level of B_1 when $\omega_1 \geqslant \Delta\omega$ vector $\mathbf{M}_1$ precesses around vector $\boldsymbol{\omega}_1$ as before, however, magnetic moment $\mathbf{M}_2$ now goes around vector $\boldsymbol{\omega}_1 + \Delta\boldsymbol{\omega}$ with a frequency $(\omega_1^2 + \Delta\omega^2)^{1/2}$. Thus the singlet–triplet mixing now proceeds with new frequency:

$$\Omega = (\omega_1^2 + \Delta\omega^2)^{1/2} - \omega_1 \cong \Delta\omega^2/2\omega_1$$

that is less than $\Delta\omega$ (Fig. 5.6b).

Thus at a high enough amplitude of an a.c. resonant magnetic field, its action not only induces transitions between T_0 and T_{+1}, T_{-1} levels but also makes the connection between S and T_0 states weaker. While studying the RYDMR spectrum by the yield of products in the singlet channel of reaction (e.g., by monitoring the intensity of fluorescence produced by recombination of ion radical pairs) one can see that the amplitude of the spectrum diminishes with an increase in the power of microwaves and even changes its sign. Such a behavior is accompanied by broadening of components of the spectrum, which is caused by an increase of the frequency of microwave-induced transitions between states. The phenomenon described is called as spin locking.

An exchange interaction, which causes a splitting between singlet and triplet levels of the pair, gives similar effects of reversing the RYDMR spectra at high enough microwave power. The dependence of the amplitude of RYDMR spectra (which are measured by scanning the field B_0) on B_1 (so called B_1 spectra) provides information about an interaction in the pairs. Moreover as the field B_1, when it is high enough, influences the mixing of all the magnetic states of the pair including the singlet one, a possibility appears of studying kinetic parameters of the pair. Indeed, the change of the mixing frequency from $\Delta\omega$ to $\Delta\omega^2/\omega_1$ can influence the properties of the mixed ST_0 state only in the case that such a state has a lifetime $\tau_{pair} < \omega_1/\Delta\omega^2$. If $\tau_{pair} \gg \omega_1/\Delta\omega^2$ the degree of mixing and, hence, properties of the pair will not depend on the frequency of mixing. By varying ω_1 ($\omega_1 \sim B_1$) within broad limits one may satisfy the first inequality. Thus studying B_1 spectra provides the possibility of determining both $\Delta\omega$ and τ_{pair}.

5.3. On the Theory of RYDMR Spectra

For theoretical description of the RYDMR spectra (as dependent both on B_0 and B_1) it is necessary to consider the spin evolution of the pair under the action of external magnetic field, Zeeman, and hyperfine interactions, to take into account a motion of the particles in the media, and the possibility of spin-dependent reactions when particles approach each other. In the general form the description requires the solution of the stochastic Liouville equation for density matrix ρ for electron spins S_i, nuclear spins I_j, distances r between particles within the pair, and orientation of the vector connecting them, Ω. Following Kubo [11] one can write the general Liouville equation

$$\rho(\mathbf{r}, \Omega, t) = -i[\hat{H}(\mathbf{r}, \Omega), \rho(\mathbf{r}, \Omega, t)] + \Gamma(\mathbf{r}, \Omega)\rho(\mathbf{r}, \Omega, t) + K(\mathbf{r}, \Omega) \tag{5.9}$$

It has three terms in the right part. The first one describes coherent spin motion under the action of Hamiltonian $H(\mathbf{r}, \Omega)$, the second term describes the motion of particles and the orientation of the pair, and the third term deals with reactions. The Hamiltonian $\hat{H}(\mathbf{r}, \Omega)$ in general form includes all the interactions that govern spin evolution

$$\frac{1}{\hbar}\hat{H} = \frac{g_1\beta B_0}{\hbar} S_1^z + \frac{g_2\beta B_0}{\hbar} S_2^z + J(S_1^z S_2^z) + \sum A_n S_1^z I_n + \sum A_k S_2^z I_k \tag{5.10}$$

$$+ \frac{g_1\beta B_1}{2\hbar}(S_1^+ e^{-i\omega t} + S_1^- e^{-i\omega t}) + \frac{g_2\beta B_1}{2\hbar}(S_2^+ e^{-i\omega t} + S_2^- e^{-i\omega t})$$

Here B_0 and B_1 are amplitudes of permanent and a.c. magnetic fields, J is an exchange interaction, ω is the frequency of the a.c. magnetic field, and $A_{n,k}$ are hyperfine interaction constants for n nuclei of the first particle and k nuclei of the second particle of the pair. One can express the probabilities of reactions through singlet and triplet channels using the density matrix:

$$w_S = k_s \int_0^\infty \mathrm{Tr}[P_s\rho(t)\,dt] \quad \text{and} \quad w_T = k_T \int_0^\infty \mathrm{Tr}[P_T\rho(t)\,dt] \tag{5.11}$$

Here P_S and P_T are projection operators for singlet and triplet states, respectively.

The solution of the Eqs. (5.9) and (5.10) in the general form has many difficulties. A review of the situation in the field is given in Steiner and Ulrich [3]. However, many important cases may be treated analytically or by using computer simulation.

The first theoretical work permitting estimation of the value of B_1, which is necessary for observing the effect of radiowaves on the rate of recombination of radicals in liquid solution, was done by Kubarev and Pshenichnov in 1974 [10]. The authors studied the cage recombination of a pair of radicals with different g-factors and nonmagnetic nuclei. Using the kinetic equation of Johnson and Merrifield [12], they showed that a high-frequency magnetic field might resonantly accelerate the recombination of a radical pair. Evaluations were made assuming that the values $\Delta g = g_1 - g_2$ and B_0 were such that the spacing between each pair of magnetic level is very large compared to their width. In other words, the nondiagonal elements of the density matrix were assumed to be negligible. It was also assumed that a high-frequency field affects only one kind of radical in a pair, that is, that $B_0|\Delta g| \gg gB_1$. Under these conditions Kubarev and Pshenichnov derived an expression for the recombination rate constant. Since their primary purpose was to clarify the conditions in which a high-frequency field would have a large effect on a process, their quantitative evaluations were not very consoling: $B_0 \cong 10^5$ to 10^6 G and $B_1 \cong 10$ G, with a radical pair lifetime of $\tau \cong 10^{-7}$ s. Later, however, it was shown that even slight variations in the recombination probabilities possible in fields about 10 to 15 times as weak as those mentioned might be discernible in RYDMR spectra. Kubarev et al. [13] calculated the probabilities of the geminate recombination of radicals for two cage models, a static and dynamic one. Analytical expressions for these probabilities were derived on the assumption that a recombination probability is mainly due to the first repeated contact which is quite justified for singlet precursors. For the sake of simplicity, pair-forming radicals were not assumed to contain magnetic nuclei. The analytical expression for the recombination probability obtained [13] prior to experimental observation of RYDMR spectra gives the essential shape of this spectrum.

A relatively simple case appears when one can consider the motion of particles, the spin evolution, and reactions separately. The so-called exponential or cage model treats the pair of paramagnetic species as a single system that decays and reacts by first-order kinetics. Phenomenological rate constants are usually used. For pairs with a long enough lifetime a quasi-steady state is established that can be described by a number of states whose population densities may be calculated by the solution of the system of kinetic equations. The microwave field B_1 induces transitions between the states. In such an approach the RYDMR spectra (as a function of B_0) may be described as a simple superposition of EPR spectra of particles of the pair with broadening caused by transition frequency. But such an approach is unable to describe effects connected with the influence of B_1 on mixing of spin states, and the RYDMR B_1 spectra cannot be calculated.

A more general approach to calculation of RYDMR spectra for radical pairs makes use of the solution of the stochastic Liouville equation in the rotating frame. Within a so-called secular approximation, which is valid for small hyperfine interaction in comparison with Zeeman one, the solution for the exponential model may be obtained in the analytical form. Lersch and Michel-Beyerle [14] obtained the formulas that are able to describe RYDMR spectra (as dependent on B_0 and B_1) for pairs that have a permanent exchange interaction. They may react with rate constants k_S and k_T through singlet and triplet channels but have negligibly small dipole–dipole interaction. A useful formula was obtained [14] for the intensity of the RYDMR spectra $F(B_0, B_1) = \varphi_T(B_0, B_1)/\varphi_T(B_0, 0)$, monitored by the yield φ_T of products of the triplet channel, which describes both B_0 and B_1 spectra well:

$$F = 1 + \frac{1}{2}\frac{B_1^2}{B_{\text{eff}}^2}\left[\frac{J^2 + k^2}{(J + g\beta B_{\text{eff}})^2 + k^2} + \frac{J^2 + k^2}{(J - g\beta B_{\text{eff}})^2 + k^2} - 2\right] \tag{5.12}$$

where $B_{\text{eff}} = \{B_1^2 - [B_0 - (\hbar\omega/g\beta)]^2\}^{1/2}$, $\quad k = \frac{1}{2}(k_S + k_T)$

Figure 8.2 shows B_0 spectra calculated on the basis of formula (5.12) at different values of B_1. Experimental data for pairs of ion radicals in the reaction centers of photosynthesizing systems may be described satisfactory by formula (5.12), thus permitting parameters J and k_S, k_T to be calculated (see Chapter 8).

5.4. Parameters Depending on the Rate of Processes in Pairs

RYDMR spectra may be obtained by monitoring any parameters of the chemical systems under study, but it must be connected kinetically with the population of magnetic sublevels of the pair. Reaction products in pairs consisting of particles with a spin $s = 1/2\hbar$ (in doublet–doublet pairs) are produced generally speaking in three channels: in the singlet (rate constant k_S), in the triplet (k_T), and in the dissociation channel (k_{-1}), the latter having a rate that does not depend on the spin. In triplet–doublet pairs the same occurs in the doublet, quadruplet, and dissociation channels. In triplet–triplet pairs it occurs in the singlet, triplet, quintuplet, and dissociation channels. Among the most convenient parameters for measuring the RYDMR signal the following.

The Intensity of Luminescence

Electronically excited molecules that are able to fluoresce (if appeared in the singlet channel) or phosphoresce (in the triplet channel) are the products of reactions in pairs. It occurs, for example, at the recombination of ion radicals D^+ and A^- in liquid or solid solutions, and in molecular crystals:

$$^1(D^+ + A^-) \Rightarrow {}^1D_1 + A_0 \Rightarrow D_0 + A_0 + h\nu \text{ (fluorescence)} \quad (5.13)$$

$$^3(D^+ + A^-) \Rightarrow {}^3D^+ + A_0 \Rightarrow D_0 + A_0 + h\nu \text{ (phosphorescence)} \quad (5.14)$$

By monitoring fluorescence the processes of spin evolution and production of products may be resolved in time. The modulation technique may be used, which permits higher sensitivity by measuring the signal on the frequency of the modulation of microwaves and using a lock-in registration of the signal.

Let us estimate some conditions of the reaction that can be studied by the RYDMR method. Let the reaction take place in the sample within a volume $V \cong 0.1\ \text{cm}^3$, and the quantum yield of the fluorescence (the number of quanta emitted per one act of recombination) $\phi \cong 10^{-2}$, an efficiency of collection of the light from the sample $\theta \cong 10^{-2}$, and the bandwidth of a detection system $\Delta f \cong 1\ \text{s}^{-1}$. The noise of the light is usually a shot noise by its nature, which is connected with quantum nature of the light. Then the connection between the rate g of the production of pairs that is proportional to the rate of the reaction (5.13), and the relative accuracy r of the measurement of the signal is given by

$$r = [\Delta f/(g\varphi\theta v)]^{1/2} \quad (5.15)$$

By using the typical values of parameters given above we come to an accuracy $r \cong 1\%$ at the rate of pair production $g \cong 10^9$ 1/cm^3 s. Lifetimes τ_{pair} of pairs of paramagnetic species are typically within the range from 10^{-9} to 10^{-7} s. It follows that the steady-state amount of pairs in the sample that can give the RYDMR signal detected by the fluorescence with an accuracy of 1% is only $N = g\tau_{pair}v \cong 10^2$. (It is worth comparing this value with a minimum amount of paramagnetic species that can give a signal in ordinary EPR technique: 10^{11} spin/sample).

When they study reactions in pairs of triplet particles a delayed fluorescence produced by the annihilation of triplet–triplet pairs is often used as a parameter to be monitored. The intensity of phosphorescence as a rule is small due to nonradiative quenching of triplet excited states, and therefore it is rarely used, though measurements at low temperature are possible. Saik et al. [7] were able to monitor the sensitized phosphorescence of diacetyl molecules to get RYDMR spectra in a liquid solution.

Optical Absorption

Triplet molecules that are produced by recombination of ion radicals can have a lifetime large enough for their accumulation in a measurable amount. Triplet–triplet absorption of the probing light is then used for their monitoring. To create an optical density of $D_{min} \cong 10^{-2}$ on the path of the probing light through the sample of 1 cm at a typical lifetime of triplets $\tau_T \cong 10^{-4}$ s one needs to have the rate of production and recombination of pairs about $g \cong 10^{18}$ 1/cm^3 s. This value is quite achievable at pulse excitation.

The absorption of the probing light by other products of the reaction in pairs is also acceptable and possible.

Electrical Conductivity

Dissociation of pairs produces free paramagnetic particles. If they are charged one can measure an induced electrical conductivity that is proportional to the density of free charge carriers. The rate of generation of free charge carriers is proportional to the sum of the population densities of pairs in all the spin states, $G = k_{-1} \Sigma n_i$. A universal estimation of the sensitivity of the technique based on the induced conductivity $\sigma = G\tau\mu e$ can hardly be made due to uncertainty of the lifetimes τ of free carriers and of their mobilities μ. In the limit of the most favorable but hardly achievable case when all free charges are able to get to electrodes, the current measured gives the same sensitivity as estimated for the luminescence mode of registration.

The Concentration of Free Radicals

Dissociation of pairs of neutral free radicals in liquid solutions leads to their appearance in the volume of active particles that are able to interact with some additives that may be introduced into the solution intentionally. Stable products may be produced as a result of such interaction, the amount of which may be measured differently. So, in Okazaki et al. [15], for example 3,5-dibromo-4-nitrosobenzenesulfonate or phenyl-*t*-butylnitron, as spin traps, were added to the liquid micellar solution where free radicals were produced by photolysis of menadione. Stable trapped radicals were accumulated in the sample within the cavity of the RYDMR spectrometer during the time of the excitation of the solution in the presence of an acting resonant microwave field (within minutes). To determine the amount of the radicals accumulated the magnetic spectrometer was then switched over to an ordinary mode of operation when the probing microwave power absorption was measured. Grant et al. [20] used the probing beam of the light to excite the fluorescence of free radicals escaped from pairs to the bulk of the solution. The intensity of fluorescence was a measure of the amount of free radicals.

Nuclear Polarization

Dynamically induced polarization of spins of radicals due to a hyperfine interaction gives rise to a nuclear polarization of diamagnetic products of spin-dependent reactions. It provides the possibility of measuring RYDMR spectra by the amplitude of the signal of chemically induced dynamic nuclear polarization (CIDNP). Diamagnetic products of reactions in pairs generated in a sample within the cavity of RYDMR spectrometer were transferred in a continuous flow to the nuclear magnetic resonance spectrometer in a time interval shorter than the spin-lattice nuclear relaxation time (<1 s) [16].

5.5. RYDMR Spectroscopy Technique

5.5.1. Resonating Cavities and Spectrometers

Spectroscopy of RYDMR makes use of the same transitions between electron magnetic sublevels as does ordinary EPR spectroscopy. The difference consists only in the level of microwave power. An estimation made above (see Section 5.2.3) shows that a high enough signal detected by changes of the fluorescence intensity is obtained by applying the microwave magnetic field $B_1 \cong 1$ G. For cylindrical cavity with TE 011 type of oscillations, a connection between the value of B_1 and a microwave power P entering into the cavity with a quality Q is given by (from [17]):

$$B_1 = 4.5 \times 10^{-2}\sqrt{QP} \tag{5.16}$$

At the quality $Q = 10^3$ and $B_1 = 1$ G, $P \cong 1$ W. That value is one or two orders of magnitude higher than the power used in ordinary EPR spectrometers. In a continuous mode of operation klystrons or Gann diodes are usually used as a source of microwaves, and some amplification of the power is provided by a traveling wave tube. To get B_1 spectra one needs still higher values of B_1 up to 100 or even 200 G, but their action is acceptable in the pulse regime only. The cavity of an RYDMR spectrometer must be adequately arranged to permit carrying out the reaction under study in the region of localization of the B_1 field. Often reactions are initiated by light or ionizing radiation, so the cavity must have proper windows for entering of the exciting beam and exiting of the detection of the signal.

A simple and convenient attachment to an EPR spectrometer is shown in Fig. 5.7. It permits RYDMR spectra to be obtained from liquid solutions monitoring the luminescence excited by X-rays. The attachment was used to obtain the first RYDMR spectra in liquids [19]. A X-ray tube, quartz light guide, and photomultiplier connected with the input of the lock-in detection system were used. The shield of the light guide was used to supply a cooled N_2 gas, thus permitting it to work at temperatures down to 100 K. The dose rate of the irradiation in experiments was about 85 krad/h; microwave power equaled 300 mW.

A cell for studying RYDMR spectra of intermediate pairs of radicals produced in photo- and thermochemical reactions by monitoring the absorp-

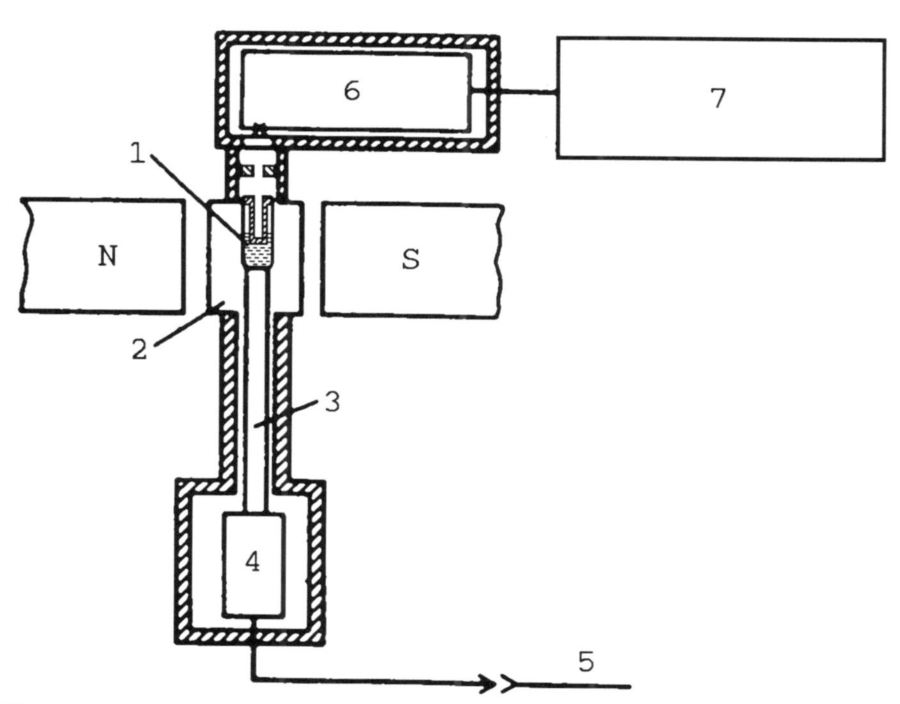

Figure 5.7 A scheme of an attachment to a standard EPR spectrometer permitting an optical detection of the RYDMR spectra. An X-ray tube is used as a source of ionizing radiation. (1) Sample, (2) cavity, (3) light guide, (4) photomultiplier, (5) exit, (6) X-ray tube, (7) power supply for the X-ray tube. From Anisimov [18].

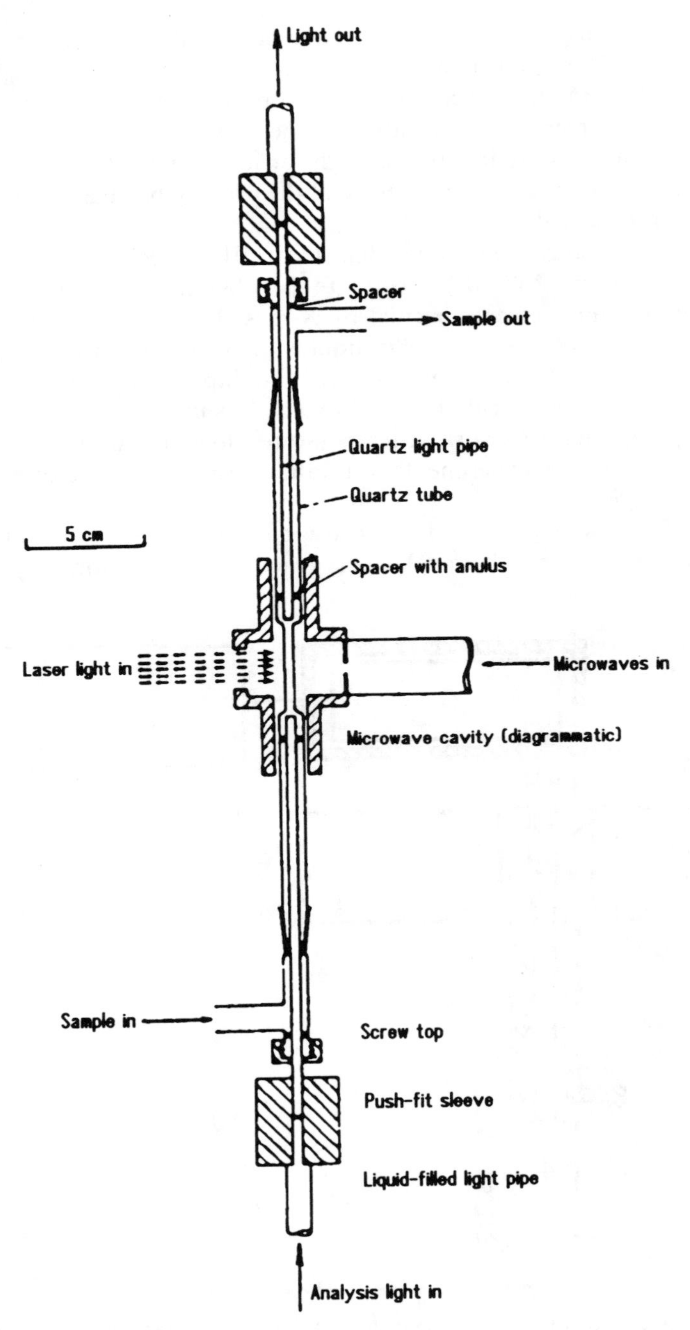

Figure 5.8 A scheme of the quartz cell for investigation of the liquid solutions by the RYDMR method. From Grant et al. [20].

tion spectra of reaction products is shown in Fig. 5.8 [20]. Radicals were produced inside the thin flow reactor situated along the axis of a cylindrical cavity of the TE 102 type. Polar liquid solutions could be used. Quartz light guides are attached to the ends of the flow reactor tube. The reactor can be enlightened by probing light with continuous spectra of a 150-W incandescent lamp. A laser provided an excitation light that entered the reactor on one side. A block scheme of a spectrometer where such a cell was used is shown in Fig. 5.9.

To avoid high electrical losses in the cavity caused by absorption of microwave power by polar liquid it is useful to work with the lowest possible frequencies of radiowaves. This means that a low magnetic field B_0 must be used to provide the resonant conditions. An RYDMR spectrometer is described [21] where X-ray excited liquid polar samples may be investigated by monitoring the recombination luminescence. Radio-frequencies from 30 to

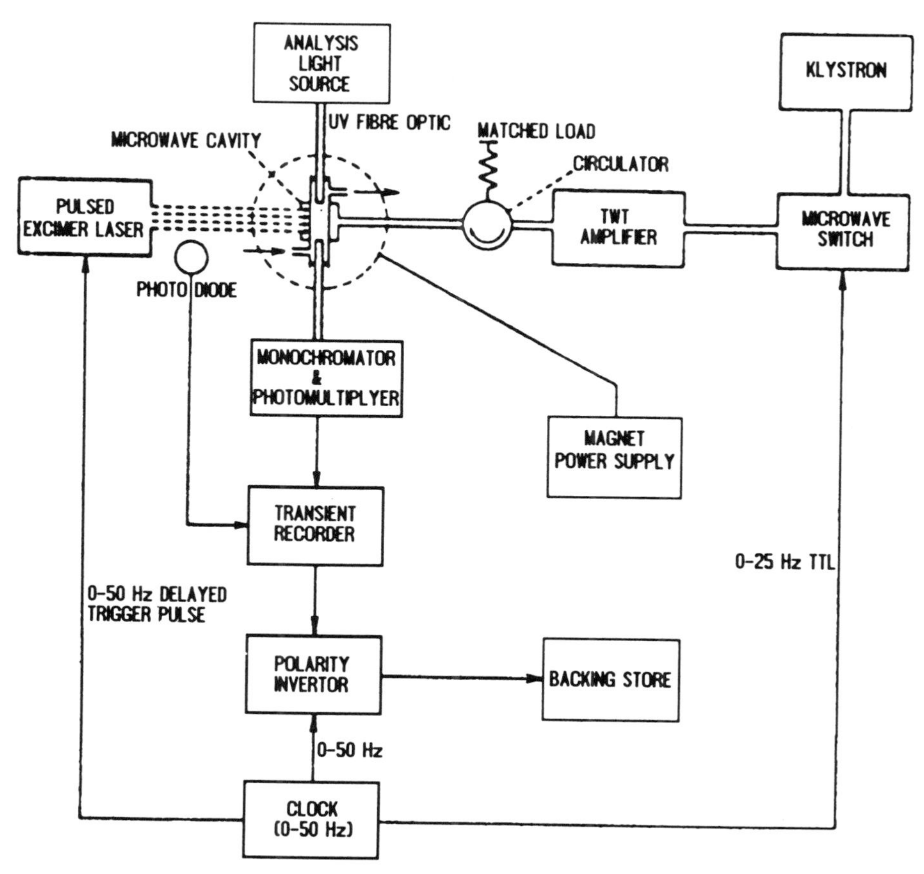

Figure 5.9 Simplified scheme of the RYDMR spectrometer. From Grant et al. [20].

500 MHz and B_0 up to 250 G were used in the spectrometer. The probe of the low-field spectrometer is shown schematically in Fig. 5.10. A 1-cm^3 sample in a quartz tube was inserted into an LC circuit coil and irradiated continuously by an X-ray tube through the gaps between coil loops. The light was transmitted from the sample through a quartz light guide to a screened photomultiplier tube. Optical detection of the spectra is achieved by a phase-sensitive detection system.

For studying photochemical reactions in liquid solutions a loop-gap resonator is convenient. By its high filling factor, its excellent field homogeneity, and its favorable conversion of microwave power into microwave field strength the loop-gap resonator is especially suited for high-power EPR and RYDMR investigation of dilute biological samples. Lersch et al. [22] used a two-loop, two-gap resonator working in S band (2 to 4 GHz). This frequency band was chosen because of the lower microwave absorption losses for aqueous samples as compared with the X band ($\cong$ 9 GHZ). The resonator is shown in Fig. 5.11. It is made of copper, has an inner diameter of 10 mm, a gap width of 1.4 mm, an outer diameter of 18 mm, and a length of 15 mm. To avoid radiation it is surrounded by a cylindrical shield. A cylindrical cuvette shown in Fig. 5.12 fitted the inner size of the quartz tube inside of the resonator. The sample was situated between two inner surfaces of the light pipe with a diameter of 5 mm.

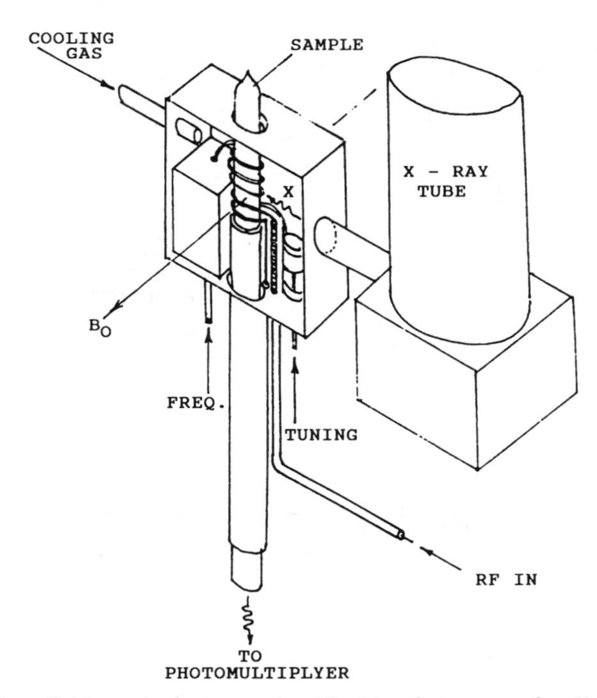

Figure 5.10 Low-field spectrometer probe. Working frequency of radiowaves is from 30 to 500 MHz; magnetic field B_0 is up to 250 G. From Koptyug et al. [21].

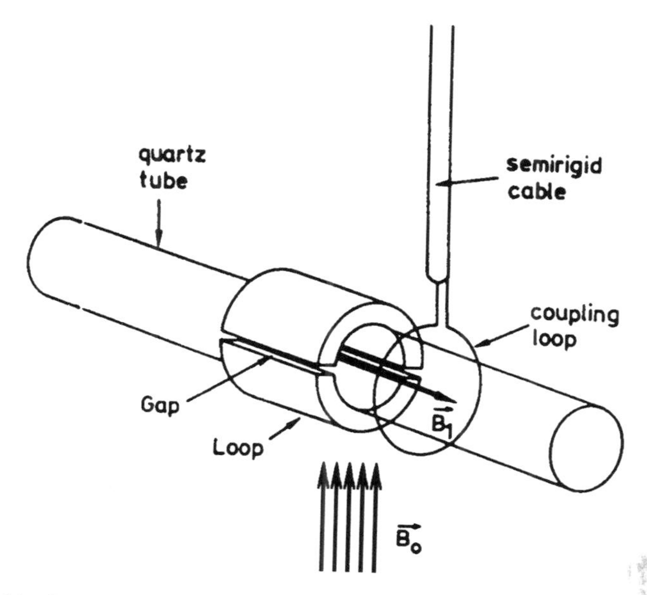

Figure 5.11 Loop gap resonator with coupling loop. From Lersch et al. [22].

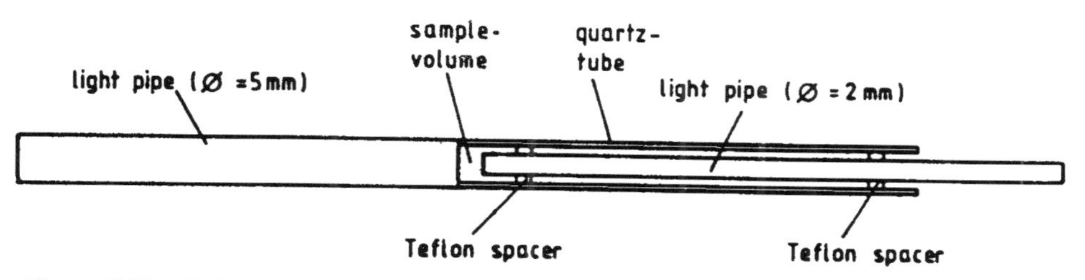

Figure 5.12 Cylindrical cuvette permitting longitudinal optical excitation and detection with the sample inside a loop-gap resonator. From Lersch et al. [22].

It was possible to achieve $B_1 = 35\,\mathrm{G}$ using microwave power $P = 1\,\mathrm{kW}$ at Q value about 400 ∓ 100 for the water sample.

An interesting technique of monitoring the reaction products in liquid solutions has been described [23]. It consisted of measuring the amount of free radicals escaped from micelles. Spin traps added to the solution converted active radicals into stable ones, the amount of which was measured in the same cavity after finishing the period of their production under the action of light and under the influence of B_0 and B_1 fields.

Figure 5.13 shows a block scheme of a spectrometer where the RYDMR principle was used to study the processes of generation and recombination of charge carriers in polymeric organic semiconductors [24]. Samples were excited by the light of an Ar^+ laser. Intermediate species that were produced

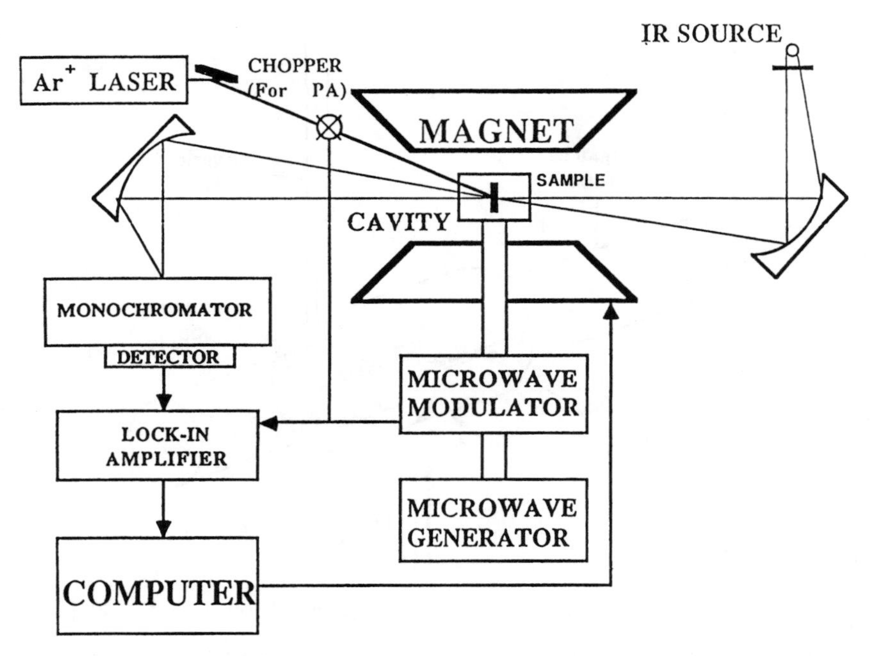

Figure 5.13 A scheme of the magnetic spectrometer used to study intermediate species in semiconducting polymers. From Wei et al. [24].

in the processes of recombination in pairs of paramagnetic species were monitored by absorption of the probing light in the IR region of the spectra. A modulation of microwave power was used, and the signal was detected by the lock-in system.

When photoconductivity is used as a technique of detection of the exit signal, a two electrode technique is usually used. Electrodes are applied to the sample situated in the region of high B_1 inside the cavity. An a.c. component of the photocurrent through the sample is measured while d.c. voltage is applied to electrodes, and microwave power is modulated [25].

5.5.2. Pulse Techniques

Pairs of paramagnetic species are produced as a rule in a "pure" spin state, which is the same as their precursors had. The production is followed by spin evolution and by different reactions in the pairs. Pulse techniques of RYDMR spectroscopy are being used to study the kinetics of these processes. Moreover use of pulse microwave power permits us to superimpose the B_1 field of high intensity (up to 200 G) and to escape the danger of heating the sample by the microwaves absorbed. That danger prevents applying continuous microwaves with power higher than 1 W ($B_1 = 1$ G).

The generation of pairs is usually provided by a pulse of light or ionizing radiation of nanosecond duration. A microwave pulse is applied afterward and it acts during the time interval, which is long enough to mix singlet and triplet states of the pair (up to microseconds). The detection of the signal is done within the time window that is open during the time determined by the kinetics of the reaction (up to milliseconds). It is possible, in principle, to vary the positions and duration of pulses and window. The effect of microwaves on the yield of products is usually small ($<1\%$)and therefore a differential technique is often used to extract a small change in the yield caused by microwaves. In Grant et al. [20] that was achieved by using the sequence of pulses shown in Fig. 5.14. Radicals in a liquid solution were generated by pulses of an eximer laser, which provided pulses of 10 to 20 ns duration on a wavelength of 308 nm with a repetition rate of 20 Hz. Any of the substances directly formed, or products of their reactions, were monitored in the time following the flash by detecting optical absorption spectrum, at a given wavelength, to yield a decay curve that was stored in a computer. When the next flash occurred it was with the application of a microwave field, so that the concentration being measured was changed. The resulting decay curve was subtracted from the one obtained from the first flash in the computer whose store then contained only the difference of the two. In this way results obtained with the field on and off were compared during ca. 20 ms for each cycle. Averaging during a big number of cycles permits the pure signal caused by microwaves to be set.

The position of pulses for generation, application of microwaves, and registration of window in respect to a typical kinetic curve for decay of

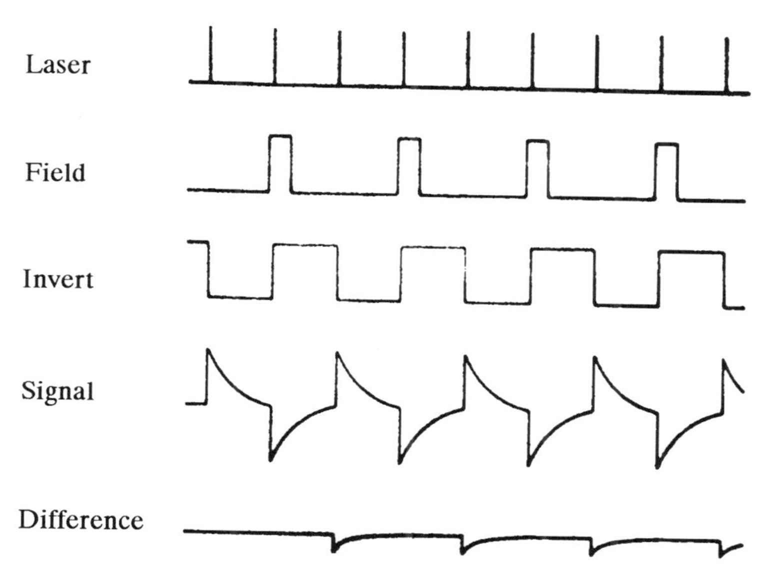

Figure 5.14 A sequence of pulses for investigation of reactions in pairs by the RYDMR method. From Grant et al. [20].

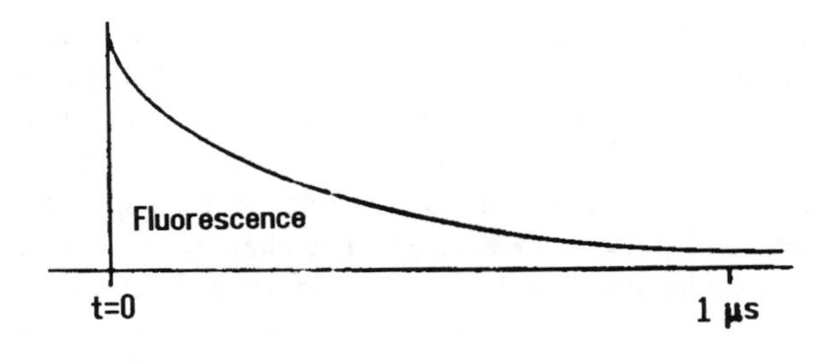

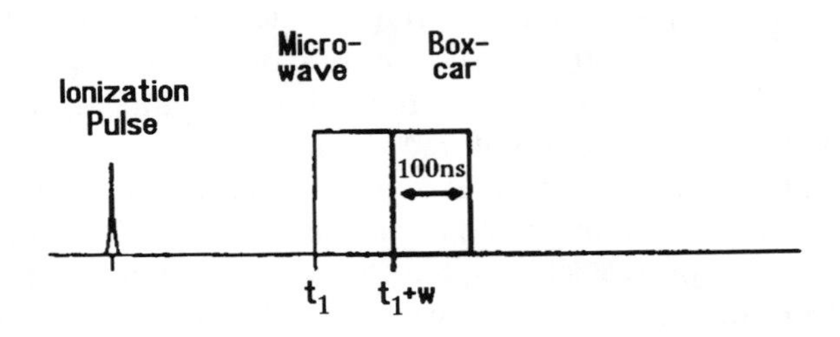

Figure 5.15 Positions of pulses that determine the generation of pairs, an action of microwaves, and the registration of a signal. From Werst and Trifunac [26].

recombination fluorescence is shown in Fig. 5.15. Such a pulse system was used in works studying the reactions in liquid solutions of organic substances initiated by the action of fast electrons [26].

5.6. Main Conditions of the Existence of the Magnetic Resonant Effect on the Reaction Rate

In conclusion we shall mention the principal conditions that must be satisfied for proper application of the RYDMR method to study the mechanism of reactions:

1. A reaction must involve a stage of production of a pair of paramagnetic species that are able to interact with each other at least in two competitive channels, which lead to different reaction products. The choice of the channel must depend on the total spin of the pair.
2. Pairs must be polarized: population densities of spin sublevels of the pair must differ due to either different rates of population or nonequal reaction rates in the channels.

3. As resonant transitions induced by electromagnetic power absorption are possible only between the levels of the same multiplicity, a participation in the transition of at least one level of mixed multiplicity is necessary to change the spin state of the pair.
4. The lifetime of the pair must be sufficiently long to permit changing the spin state of the pair under the action of the magnetic field of an electromagnetic wave.
5. The intensity of the signal connected with the reaction rate must be high enough and stable in time permitting a measurement of small changes induced by radiowaves on the background of the total signal.

References

1. Clarke Richard, H., ed. *Triplet State ODMR Spectroscopy*. Wiley-Interscience, New York, 1982.
2. Leung, M., El-Sayed, M. A. *J. Amer. Chem. Soc.* **1975**, *97*, 669–674.
3. Steiner, U. E., Ulrich, T. *Chem. Rev.* **1989**, *89*, 51–147.
4. Salikhov, K. M., Molin, Yu. N., Sagdeev, R. Z., Buchachenko, A. L. *Spin Polarization and Magnetic Effects in Radical Reactions*. Elsevier, Amsterdam, 1984.
5. Zeldovich, J. B., Buchachenko, A. L., Frankevich, E. L. *Soviet Phys.—Uspekhi* **1988**, *31*, 385–408.
6. Frankevich, E. L., Kubarev, S. I. Spectroscopy of reaction yield detected magnetic resonance. In *Triplet State ODMR Spectroscopy*, Clarke, R. H., ed. Wiley-Interscience, New York, 1982, pp. 138–184.
7. Saik, V. O., Anisimov, O. A., Molin, Yu. N. *Chem. Phys. Lett.* **1985**, *116*, 138–142.
8. Frankevich, E. L., Pristupa, A. I., Lesin, V. I. *Chem. Phys. Lett.* **1978**, *47*, 304–309.
9. Frankevich, E. L. *Zh. Eksper. Teor. Fiz.* **1966**, *50*, 122–126. English translation: *Soviet Phys. JETP* **1966**, *23*, 814–820.
10. Kubarev, S. I., Pshenichnov, E. A. *Chem. Phys. Lett.* **1974**, *28*, 66–70.
11. Kubo, R. *J. Phys. Soc. Jpn.* **1969**, *26* (Suppl.), 1.
12. Johnson, R. C., Merrifield, R. E. *Phys. Rev.* **1970**, *B1*, 816.
13. Kubarev, S. I., Pshhenichnov, E. A., Shustov, A. S. *Teor. Eksper. Khem.* **1976**, *12*, 397. In Russian.
14. Lersch, W., Michel-Beyerle, M. E. *Chem. Phys. Lett.* **1987**, *136*, 345–350.
15. Okazaki, M., Shiga, T., Sakata, S., Kona, R., Toriyama, K. *J. Phys. Chem.* **1988**, *92*, 1402–1404.
16. Bagryanskaya, E. G., Grishin, Yu. A., Sagdeev, R. Z., Leshina,, T. V., Polyakov, N. E., Molin, Yu. N. *Chem. Phys. Lett.* **1985**, *117*, 220–224.
17. Pool, C. P. *Electron Spin Resonance*. Wiley, New York, 1967.
18. Anisimov, O. A. *J. Indust. Irradiation Tech.* **1984**, *2*, 271–300.
19. Molin, Yu. N., Anisimov, O. A., Grigoryants, V. M., Molchanov, V. K., Salikhov, K. M. *J. Phys. Chem.* **1980**, *84*, 1853–1856.

20. Grant, A. I., McLauchlan, K. A., Nattrass, S. R. *Mol. Phys.* **1985**, *55*, 557–569.

21. Koptyug, A. I., Saik, V. O., Anisimov, O. A., Molin, Yu. N. *USSR Acad. Sci. Trans.* **1987**, *297*, 1414–1419.

22. Lersch, W., Lendzian, F., Lang, E., Feick, R., Möbius, K., Michel-Beyerle, M. E. *J. Magn. Res.* **1989**, *82*, 143–149.

23. Okazaki, M., Sakata, S., Konaka, R., Shiga, T. *J. Chem. Phys.* **1987**, *86*, 6792–6800.

24. Wei, X., Hess, B. C., Vardeny, Z. V. *Phys. Rev. Lett.* **1992**, *68*, 666–670.

25. Frankevich, E. L., Tribel, M. M., Sokolik, I. A., Pristupa, A. I. *Phys. Stat. Sol. (b)* **1987**, *87*, 373–381.

26. Werst, D. W., Trifunac, A. D. *J. Phys. Chem.* **1991**, *95*, 3466–3477.

CHAPTER

6

RYDMR in Solids

6.1. Paramagnetic Species in Molecular Solids

Reactions in molecular solids that are often spin dependent are mainly connected with processes of electron or excitation transfer. Reactions that include the transfer of atomic species even if they do occur proceed at a low rate, and conditions of the conservation of spin during the course of the reaction are not satisfied.

6.1.1. Electronic Processes in Molecular Solids

These are initiated by the action of light or ionizing radiation on a substance. Studying the mechanism of the processes of radiation energy conversion gives valuable information not only for molecular physics but also for such branches of science as biology, radiation physics and chemistry, and photochemistry. The most widely studied phenomena are the optical and photoelectric properties of molecular crystals.

Typical substances for molecular crystals are aromatic compounds such as anthracene and tetracene that occupy a place in the physics of organic solids analogous to that of germanium and silicon in the physics of inorganic semiconductors. Figure 6.1 shows a diagram of energy levels of the excited and ionized states of anthracene and tetracene. We shall describe the main electronic processes that are initiated by external excitation. Light absorption gives rise to singlet molecular excitations called molecular excitons (or Frenkel excitons) that can migrate through the crystal. Singlet excitons are deactivated

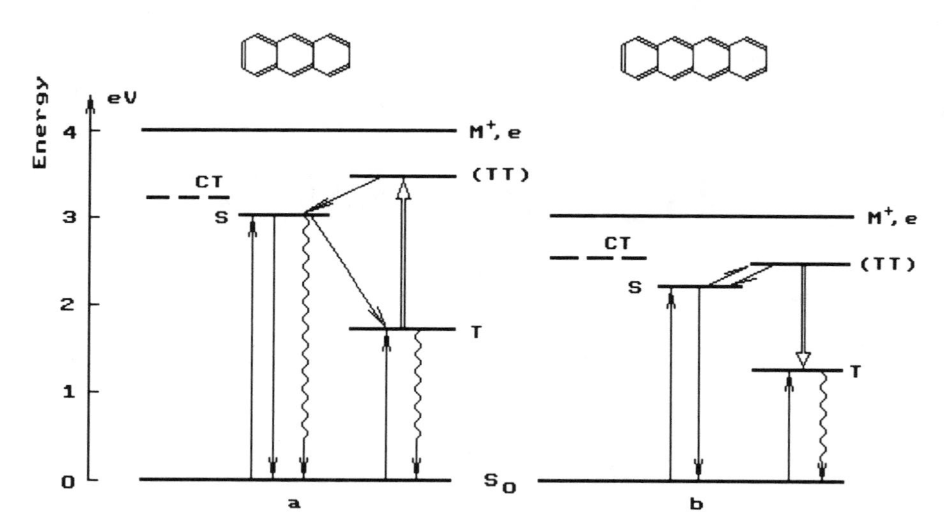

Figure 6.1 Schematic diagram of energy levels for crystals of anthracene (a) and tetracene (b). S_0 is the ground state of the crystal; S and T are energy levels of singlet and triplet excitons; (TT) is an energy state formed upon collision of two triplet excitons; CT is the lowest level of charge-transfer states, their energy position not being defined exactly; M^+, e is the ionized state of the crystal; transitions to it result from interactions of excited states with one another and/or with impurities; these transitions are not shown in the diagram. The straight vertical arrows denote transitions between states that involve light absorption, fluorescence, and phosphorescence. The double arrows show triplet–triplet annihilation, and the wavy lines are radiationless transitions.

by emission of a quantum of light (fluorescence), by intersystem crossing to triplet molecular excitons, by interaction with impurities, by fission into pairs of triplet excitons, and by other processes. Triplet excitons can decay radiatively (phosphorescence) or without radiation, being quenched by impurities, by mutual annihilation, etc. Low-lying triplets may be excited to a higher triplet excited state by absorption of light. The lifetime for triplet excitons is several orders of magnitude larger than for singlet excitons. This defines their essential role in processes of migration and conversion of energy in molecular crystals.

Production of free charges in molecular crystals requires an energy greater than the energy of excitons. Hence, ionization under the action of the visible light usually occurs by interaction of excitons with an impurity having a low ionization potential or elevated electron affinity, and also by interaction of excitons with one another. In addition to molecular excited states in molecular crystals, there can exist so-called polar excited states, or charge-transfer states, which are formed by pairs of charges of opposite signs bound by Coulomb interaction. These states are analogous to the Wannier–Mott type of excitons in inorganic crystals. The charge-transfer states are at an energy level below the bottom of the conduction band; the position of their lowest level corresponds to localization of an electron and a hole on adjacent molecules. The probability of conversion of a molecular excited state to a charge-transfer state is determined by the overlap of the molecular orbits of adjacent molecules. An estimate of this probability gives a value of 10^{-4}. Therefore, polar excitations are practically not manifested in the absorption spectra of molecular crystals. However, their role as an intermediate stage in the formation of free electrons and holes from molecular excitons is substantial. Note that charge-transfer states can also be formed by excitation of impurity molecules (both direct, and by energy transfer from molecules of the host crystal). Here a charge of one sign proves to be localized on the impurity molecule.

A detailed review and discussion of electrophysical properties and electronic processes in molecular solids can be found in the well known book of Pope and Swenberg [1].

The most widespread methods of studying absorption and conversion of radiation energy in molecular solids are based on studying luminescence and photoconductivity. Fundamental information on the features of excitation, ionization, energy transfer, and charge carrier transport has been gained with their aid. These features involve, first, the substantial role in these processes of excited states and their interaction with impurities and with each other, and second, the low mobilities of the charge carriers.

6.1.2. An Estimation of the Mobility of Active Particles

Typical diffusion coefficients D or mobilities μ of the particles that take part in spin-dependent reactions in solids may be evaluated on the basis of the next condition: the time τ_{pair} particles stay in the pair state must be longer than the time needed for mixing of the spin states τ_{ev}, but shorter than the relaxation time T_1:

$$\tau_{ev} < \tau_{pair} < T_1 \tag{6.1}$$

The time τ_{ev} depends on the mechanism of spin evolution. For the HFI mechanism $\tau_{ev} = \hbar/(g\beta A)$, at $A = 10$ G and $\tau_{ev} = 5 \times 10^{-9}$ s; for the Δg mechanism $\tau_{ev} = \hbar/(\Delta g \beta B_0)$, at $\Delta g = 0.01$ and $B_0 = 10^3$ G, $\tau_{ev} = 10^{-8}$ s; for the dipole–dipole mechanism of spin evolution $\tau_{ev} = \hbar/(g\beta D)$, where D is the dipole–dipole energy of interaction, at $D = 10^3$ G, $\tau_{ev} = 5 \times 10^{-11}$ s.

The nature of the state of a correlated pair may be different and there are two distinct limiting cases. First, particles of the pair are close to one another throughout the lifetime of the pair so that they may react in the pair state at any instant. In this case the pair is described as a quasiparticle decaying by the first-order low, its lifetime being determined by the sum of reaction and decay rates. This may be the case when active species get stabilized in solids on traps in immediate proximity to one another or a weakly coupled complex of two species is formed. The other case corresponds to a situation in which after the first contact, active species are separated by diffusion and their new attempt to react obeys the laws of a random walk process. In the diffusion process the probability of a particle at instant t being at a distance r far from the particle with which it collided at $t = 0$ is given by the formula

$$P(r, t) = (4\pi Dt)^{-3/2} \exp(-r^2/4Dt) \tag{6.2}$$

where D is the sum of the diffusion coefficients of the two particles. From Eq. (6.2) it follows that an estimate of the time in which the particles try to make the first reencounter, that is, approach each other within the reaction radius R, is

$$\tau_{\text{pair}} \approx R^2/D \tag{6.3}$$

Thus for instance, at $R = 5$ Å and for a diffusion coefficient $D = 10^{-6}\,\text{cm}^2/\text{s}$ $\tau_{\text{pair}} \cong 2.5 \times 10^{-9}$ s. In the Coulomb interaction of charged particles forming a pair, $R \approx e^2/\varepsilon kT \approx 200$ Å and $\tau_{\text{pair}} \approx 10^{-8}$ s, even when $D = 10^{-4}\,\text{cm}^2/\text{s}$.

The probability of repeated contact depends on the relationship between the reaction radius R and the value of the diffusion jump λ. Noyes [2] demonstrated that at $R = \lambda$ the probability of repeated contact equals 0.5. At $R > \lambda$ this probability is close to unity.

The specific features of diffusion expressed by Eq. (6.2) do not allow the pairs to be characterized by an average lifetime—it proves to be infinitely long. But since at lifetimes exceeding the time of spin–lattice relaxation there are no more correlated pairs, the lifetime is practically determined by the relaxation time. A pair lifetime is then found, with due reference to the relaxation from the distribution $P(r, t)$ multiplied by the probability $S = \exp(-t/T_1)$ of preserving the initial spin correlation during a time t. Neglecting an influence of the reaction that shortens the lifetime, the average lifetime of the pair in this case is

$$\tau_{\text{pair}} = \tfrac{1}{3}(T_1 R^2/D)^{1/2} \tag{6.4}$$

Using the inequality (6.1), an estimation of evolution times made above, the reaction radius $R = 5$ Å, and relaxation time $T_1 \approx 10^{-7}$ s leads to a range of diffusion coefficients of neutral paramagnetic particles that may take part in spin-dependent reactions in pairs:

$$10^{-5} > D > 10^{-8}\ \text{cm}^2/\text{s}$$

For oppositely charged particles in molecular solids one can imagine two models of the pair. The first considers localized states of ion radicals situated on short distances from one another that can recombine in the process of electron tunneling. The condition (6.1) limits the lifetime of such pairs. The second model assumes at least one movable particle in the pair, say an electron, that can migrate in the vicinity of the positive ion on the distances $r \geqslant R_{\text{ons}}$, and thus having a high probability of geminate recombination. It follows from condition (6.1) that at $R \approx R_{\text{ons}} = 200$ Å the range of the diffusion coefficients may be estimated as $10^{-3} > D > 10^{-6}\,\text{cm}^2/\text{s}$ or the range of mobilities at room temperature as $4 \times 10^{-2} > \mu > 10^{-6}\,\text{cm}^2/\text{V s}$.

The evolution of spin of particles with spin $s = 1/2\hbar$ in organic substances is determined more often by the hyperfine interaction. However, the motion of the electron (or hole) in solids leads to averaging of that interaction to zero if the localization time of the charge on the single site is shorter than the evolution time τ_{ev}. One can expect therefore that the HFI mechanism of the mixing of spin states will be operative only for pairs in which at least one charge remains localized or it moves by hoping with a localization time $\tau \geqslant \tau_{ev}$.

Polymers are also related to molecular solids. Processes that are similar to those for molecular crystals proceed in polymers, however, they are more complicated due to the presence of less regular structures, an existence of amorphous and crystalline phases, impurities, and defects. Polymers that have a system of conjugated double bondes also have electrons delocalized through the system of conjugation and it determines their unusual electrical and optical properties. Semiconducting properties of conjugated polymers attract much attention at present. There is real hope of creating highly sensitive devices for light energy conversion. Another possible application concerns the problem of organic metals having the specific electrical conductivity of the same order as that of copper or silver. Very promising is a virtual application in optoelectronics making use of nonlinear optical properties of conjugated polymers. All these applications require the processes involving electrons, holes, triplet excitations, and other paramagnetic species to be studied. The possibility of selecting and studying the processes by using the resonant action of radiowaves offers very promising perspectives.

6.2. Spin Selectivity in Photo- and Radiation-Induced Processes

Let us consider elementary processes in molecular solids in which spin selection rules can be revealed.

6.2.1. Generation and Recombination of Charges in Molecular Crystals

The main processes accounting for the occurrence of charge carriers in molecular solids exposed to light in the singlet–singlet absorption band are interactions of singlet molecular excitons $^1D_1$ with admixtures $^1A_0$ or electrodes. As a result of the energy gain achieved upon electron localization on the admixture the energy of a singlet exciton is usually sufficient to produce free charges.

Ionization of molecules in molecular solids in any case leads to the appearance of free charges in the bulk of the crystal, although with small probability. That is due to the small free path of the electrons produced. After their production in the processes of ionization the electrons lose their initial kinetic energy and become thermalized at some distance a from the remaining positive ion. As a rule

$$a < r_{ons} = \frac{e^2}{\varepsilon kT} \tag{6.5}$$

and electrons appear in a strong electrical field that prevents their escape from the Coulomb potential well. Onsager [3] has shown that the probability of the separation of an electron-ion pair

$$p = \exp\left(-\frac{r_{ons}}{a}\right) = \exp\left(-\frac{e^2}{\varepsilon kT}\right) \tag{6.6}$$

For example at $a = 30$ Å, $\varepsilon = 3$, and $T = 300$ K, $p = 2 \times 10^{-3}$.

The transfer of an electron from $^1D_1$ to $^1A_0$ leads to a pair charge-transfer state (D^+A^-) that may either recombine or dissociate into free charges

$$\begin{array}{ccccc} & \ulcorner\!\rightarrow & D^+ + A^- & \leftarrow\!\urcorner & \\ {}^1D_1 + {}^1A_0 \rightarrow & {}^1(D^+\ A^-) & \rightleftarrows & {}^3(D^+\ A^-) & \\ & \downarrow & & \downarrow & \\ & {}^1D_0 & & {}^3D & \end{array} \tag{6.7}$$

The process (6.7) actually corresponds to the general scheme of an oxidation–reduction reaction. In molecular crystals one may consider species $^1D_1$ and D^+ to be movable as excitons and holes whereas $^1A_0$ and A^- remain localized as impurity molecule and trapped electron. According to Eq. (6.7) an initial state of the pair $(D^+\ A^-)$ is a singlet. But one can imagine that the population of the singlet and triplet states of the pair proceeds by a process of recombination of free charge carriers D^+ and A^- if they meet each other occasionally.

6.2.2. Ionization in Solid Solutions

When solid solutions are irradiated by ionizing radiation, such as X-rays or fast electrons, ion radical pairs are also produced. Organic solvent containing donor D and acceptor A admixture molecules is usually used to study the action of radiation on materials. In the process of ionization of the solvent molecules S electrons e and "holes" S^+ are produced that can be transferred to A and D:

$$S \Rightarrow S^+ + e \tag{6.8}$$

$$e + A \Rightarrow A^-$$

$$S^+ + D \Rightarrow S + D^+$$

Geminate pairs are produced of the next type: $(S^+\ e)$, $(S^+\ A^-)$, $(D^+\ e)$, $(D^+\ A^-)$.

The processes of charge transfer are fast enough to prevent the relaxation of the spin, and pairs remain in the singlet state. Molecules that are able to fluoresce are used as acceptor A and they become electronically excited in the process of the recombination of pairs. The intensity of fluorescence serves as a measure of the reaction rate in pairs.

6.2.3. Triplet–Triplet Annihilation Process

The observation of the magnetic field effect on the delayed fluorescence of molecular crystals [4, 5] was evidence of the occurrence of a spin-dependent process involving triplet excitons. Annihilation of triplet excitons is described by the next reaction:

$$^3A + {}^3A \underset{k_{-1}}{\overset{k_1}{\rightleftharpoons}} {}^{1,3,5}({}^3A\,{}^3A) \xrightarrow{k_2} {}^1A_1 + {}^1A_0 \tag{6.9}$$

The rate constant of annihilation $\gamma = k_1k_2/(k_2 + k_{-1})$ where k_1, k_{-1}, and k_2 are the rate constants of collision, back scattering, and the production of singlet products, respectively. Free excitons are not assumed to interact until the pair $({}^3A\ {}^3A)$ is produced. The lifetime of such a pair is taken to be far shorter than the spin–lattice relaxation time. Hence the reaction rate depends strongly on the spin state of the pair.

It can be shown [6] that if the near-lying levels are not degenerate, an approximate calculation based on the steady-state populations of the spin Hamiltonian of a pair, equal to the sum of the spin Hamiltonians of two free triplets, applies:

$$\hat{H} = g\beta B_0(\hat{S}_1 + \hat{S}_2) + D(S_{z1}^2 + S_{z2}^2) + E(S_{x1}^2 + S_{x2}^2 - S_{y1}^2 - S_{y2}^2) \tag{6.10}$$

A contact pair consisting of two triplets has nine spin substates, made up of a singlet, a triplet, and a quintuplet. The production rate of every substate is $1/9k_1n_T$, where n_T is the concentration of free triplet excitons. The process of scattering is independent of spin. In contrast, the spin conservation rule is essential to the annihilation process. The annihilation rate for the ith spin substate is written in the form $k_2|S_i|^2$, where S_i is the amplitude of a singlet component in the ith state. Then the annihilation probability for the ith substate is

$$\frac{k_2|S_i|^2}{k_{-1} + k_2|S_i|^2}$$

The total rate constant of annihilation is obtained by summing up the probabilities over all states:

$$\gamma = \frac{k_1}{9}\sum^{9}\frac{k_2|S_i|^2}{k_{-1} + k_2|S_i|^2} \tag{6.11}$$

It follows from Eq. (6.11) that γ increases with the number of states including a fraction of the singlet component. The cause of the magnetic field effect lies, just as in the case of doublet–doublet pairs, in that the singlet component distribution over state changes. The spin Hamiltonian in Eq. (6.10) consists of two parts describing Zeeman and zero-field splitting. At $B_0 = 0$ the eigenstates of the Hamiltonian correspond to the main values of a dipole zero-field tensor $\mathbf{H}_T$; in this case it is shown [5, 6] that in single crystal of anthracene only three spin states will contain a singlet component. Application of an external field $B_0 < \mathbf{H}_T$ will further mix the states corresponding to $\mathbf{H}_T$ and will distribute the singlet component over a greater number of states, that is, it will increase γ. Over the range of a strong field ($B_0 \gg \mathbf{H}_T$) the Zeeman splitting greatly exceeds that in a zero field, so that the latter can be taken as a perturbation in calculations. Under these conditions, spin states are quantized along an external field and only two states (i.e., less than in a zero field) exhibit singlet character. Thus, in strong fields the value of γ, and hence the fluorescence intensity, should decrease. $\gamma(B_0)$ calculated in terms of the present theory agrees fairly well with observations. When the lifetimes of triplet exciton pairs are compared with the spin–lattice relaxation time T_1, one has to take into account the dependence of the latter on the external magnetic field [7].

6.2.4. Fission of Singlet Excitons into Pairs of Triplet Ones

Variations in the rate constant of triplet annihilation due to the magnetic field affect the delayed fluorescence intensity. In the reverse process to the scheme described in Eq. (6.9), which involves the fission of a singlet excitation into a pair of triplet ones, the change of the spin state of a pair under an external magnetic field is manifested as a variation in the intensity of prompt fluorescence [8].

6.2.5. Reactions Involving Triplet–Doublet Pairs

The quenching of triplet molecular excitons by paramagnetic centers is known to reduce the intensity of delayed fluorescence and is magnetic field sensitive [9]. A collision of the triplet exciton ^{3}A with the radical ^{2}R or any other paramagnetic species with spin $S = 1/2\hbar$ may lead either to spin-independent triplet scattering or quenching. During quenching a transition is made from the initial spin state, which is a mixture of doublets and quartets, to a purely doublet final state:

$$^3\mathrm{A} + {}^2\mathrm{R} \underset{k_{-1}}{\overset{k_1}{\rightleftharpoons}} {}^{2,4}({}^3\mathrm{A}\ {}^2\mathrm{R}) \xrightarrow{k_l} {}^1\mathrm{A}_0 + {}^2\mathrm{R} \tag{6.12}$$

The rate k_l of transition from each of the six lth initial states to a final one depends on the amplitude of the doublet component therein:

$$k_l = k(|D_l^+|^2 + |D_l^-|^2) \tag{6.13}$$

where $D_l^\pm = \langle \psi_l | \psi_{\pm 1/2} \rangle$ is the amplitude of the doublet component $m_s = \pm 1/2$ in the initial state. If the scattering rate is designated as k_{-1}, the quenching probability for the initial lth substate can be written as $k_l/(k_{-1} + k_l)$, and the total rate of exciton quenching by a paramagnetic impurity will be

$$Q = \frac{A}{6} \sum \frac{k_l}{k_{-1} + k_l} \tag{6.14}$$

where A is the total rate of exciton-radical collisions. The value of Q is maximum for $|D_l^+|^2 + |D_l^-|^2 = 1/3$, for all l, that is, when the doublet component is uniformly distributed over all six substates, the minimum of Q will occur in the complete separation of doublet and quartet. Calculations have revealed that in zero field the quenching rate is at a maximum. With the application of an external magnetic field, the doublet component concentrates on four states and the value of Q decreases.

6.2.6. Hopping of Electrons Through Paramagnetic Sites

In the hopping regime of the charge transfer a spin-dependent step may occur if it happens for electron to meet a radical site $\dot{S}$. In such a case a two doublet pair "charge site" may be formed making the rate of hopping spin sensitive.

The scheme of processes describing the hopping of a "hole" h^+ is given by

$$\dot{h} + \dot{S} \rightleftharpoons {}^{1,3}(\dot{h}^+ + \dot{S}) \rightarrow h^0 + S^+ \tag{6.15}$$

The process (6.15) is allowed if during the charge hopping to a site $\dot{S}$ the multiplicity of the pair $(\dot{h}^+ \cdots \dot{S})$ coincides with the multiplicity of the charged site S^+ formed as a result of the hop. If it does not h^+ is just scattered on $\dot{S}$ not being able to be localized. Thus the frequency of hopping will be influenced by any effect on the spin state of the pair. The process (6.15) was revealed due to magnetic field modulation of the hopping conductivity of polymers [10].

6.3. Spin Selective Processes in Electron-Hole or Ion Radical Pairs

Photoconductivity of tetracene was studied using the magnetic field effect and RYDMR technique [11]. Samples of a surface and sandwich types were used. The first were 2- to 5-μm-thick tetracene films placed between two vacuum-deposited electrodes of aluminum or silver. The surface type samples were the same films deposited by vacuum sublimation on a quartz substrate having vacuum-deposited silver electrodes with a 0.2-mm-wide, 3-cm-long gap. The samples were placed in the cavity of an EPR spectrometer under a microwave magnetic field $\mathbf{B}_1 \perp \mathbf{B}_0$ and exposed to light from an incandescent lamp using glass filters transparent to the spectral region $340 < \lambda < 520$ nm. The intensity of the exciting light was up to 10^{15} quantum/cm^2; the surface area of a sample was 0.2 cm^2. At a 20 V voltage difference between the electrodes and for 300 K the photocurrent in the sandwich type sample was found to be 10^{-7} A; the same value was observed in the surface type samples at a voltage of 150 V.

Tetracene samples coated with tetraoxytetracene (TOT) were also studied. Owing to TOT coating, the value of the photocurrent increased by about an order of magnitude. The time constant of the photocurrent setup was 10^{-3} s. Variations in the magnitude of photocurrent due to a resonant microwave field were detected using a lock-in amplifier at the frequency of amplitude modulation of the microwave power.

Figure 6.2b shows an RYDMR spectrum detected by photoconductivity on a surface-type sample. The spectrum of the same shape was also detected with other samples of tetracene or TOT-coated tetracene. The spectrum consists of a single line with a halfwidth $\delta B_0 = 17$ G. The polarity is negative: photoconductivity decreases in a microwave field. An analysis of the line shape has shown that its central part is Lorentzian, with a half-width of 19 G, whereas the part 10 G or further away from the center is Gaussian, with a half-width of 25 G. Occurrence of RYDMR spectra on photoconductivity is most probably connected with pairs of charged species ($D^+ A^-$), which are produced during the free charge carriers generation. The spectrum in the form of a single line shows that in a pair state at least one charge is not localized: the frequency of reorientation of the axis coupling two charges is higher than $g\beta\delta B_0/\hbar$, that of interaction of charges in the pair. This seems to be accounted for by fast motion of the hole D^+ in the vicinity of the localized charge A^-. The tail of the Gaussian line may be attributed to the unresolved hyperfine structure of this ion. The lifetime of the pair can be evaluated from the linewidth and was found to be $\tau > 7 \times 10^{-9}$ s. In the case of the photoconductivity-detected RYDMR spectrum of the charge-transfer crystal of anthracene–tetracyanbenzene (A-TCNB) [12], a peak of different shape was revealed (see Fig. 6.2a). The peak has a flat upper part that can be decomposed into a pair of weakly resolved lines of Lorentzian shape with the magnitude of splitting between the maxima $\delta B_0 = 8$ G. The width of the signal at the halfheight is 22 G and the width of each of the component peak is $\delta B_0 = 17$ G.

One may expect that the RYDMR spectrum of the ion radical pair would consist of pairs of lines split by the dipole–dipole interaction of spins. In the spectra of polycrystalline samples the pairs of lines will be distributed with

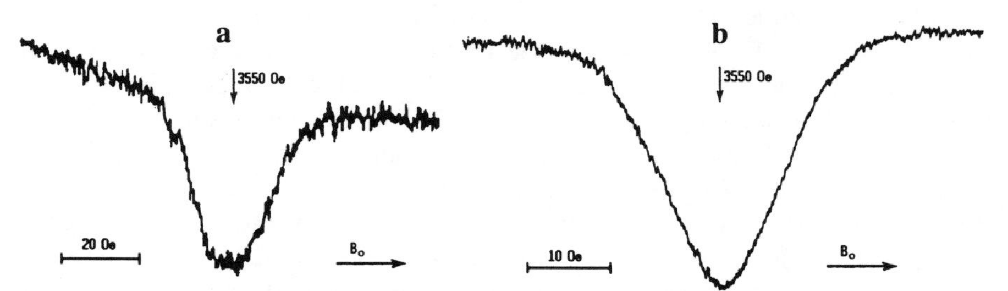

Figure 6.2 The first RYDMR spectra detected by photoconductivity at $T = 300$ K, $B_1 = 1$ G. Magnetic field B_0 sweep rate is 3 G/min. (a) For polycrystalline sample of CT complex anthracene-TCNB, (b) for tetracene film deposited by vacuum sublimation. From Frankevich et al. [11, 12].

respect to the splitting caused by the anisotropy of dipole–dipole interaction, the distance between the distribution maxima being determined by the quantity D and the total width by the quantity $2D$. The spectrum obtained can be interpreted as a pair of lines strongly broadened owing to the small lifetime of the pair, the splitting arising from the dipole–dipole interaction. It corresponds to the mean distance between the charges $r < [(3/2)(g\beta/\delta B_0)]^{1/3} \approx 15\,\text{Å}$. The width $\Delta B_0 = 17\,\text{G}$ makes it possible to estimate the lifetime of the pair as $\tau = 2\hbar/(g\beta\ \delta B_0) = 6.7 \times 10^{-9}\,\text{s}$. The nature of the pair in the CT crystal sample seems to be a charge-transfer state itself.

6.4. Spin Selective Interaction of Charge Carriers with Paramagnetic Sites

Mechanisms of the electrical conductivity of organic solids including polymers have attracted considerable attention of physicists and chemists as they permit accounting of the properties of newly developed electronic materials. Polyacetylene (PA) is a prototyped example of a material that serves as a model for mechanistic research. Polyacetylene consists of weakly coupled chains of CH units forming a pseudo-1D lattice. The stable isomer is *trans*-$(CH)_x$ in which the chain has a zigzag geometry. The *cis*-$(CH)_x$ isomer in which the chain has a slightly different backbone geometry is unstable at room temperature. *Trans*-$(CH)_x$ is a semiconductor with energy gap $E_g = 1.65\,\text{eV}$, which has two equivalent lowest energy states (L and R) having two distinct bonding structure.

Figure 6.3 shows the structure of a polymer molecule of *trans*-polyacetylene. Molecules may exist in two states that differ by the order of alternation of single and double bonds. At the point of their meeting inside the same molecule a paramagnetic site $\dot{S}$ appears. The site is capable of moving along the chain, to capture a charge, to interact with other sites $\dot{S}$. The sites behave like quasiparticles and are called solitons. In the band picture they produce an energy level situated near the center of the gap. The theory of solitons in polyacetylene was developed by Su et al. [13]. A striking feature of charged solitons that take part in the conductivity process consists of the absence of

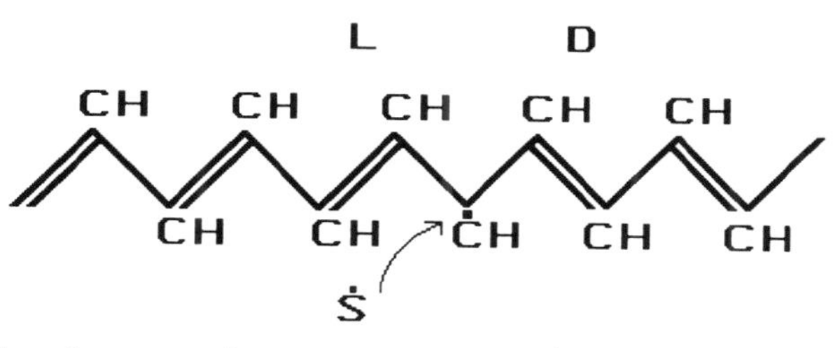

Figure 6.3 Two degenerate in energy structures of *trans*-polyacetylene, *L* and *D*. At their meeting in the same molecule a paramagnetic site $\dot{S}$ is formed.

the spin. It differentiates them sharply from ordinary charge carriers such as electrons, holes, or polarons.

The electrical conductivity of polyacetylene is determined by the presence of dopants, and, depending on their concentration, can have values between 10^{-10} and $10^{3}\,\Omega^{-1}\,cm^{-1}$. In weakly doped polyacetylene $(CH)_x$ at doping level $y \ll 10^{-3}$ ($\sigma < 10^{-5}\,\Omega^{-1}\,cm^{-1}$) a high concentration of paramagnetic defects (10^{18}–$10^{19}\,cm^{-3}$), being formed in it during polymerization and isomerization, is presented. In *trans*-$(CH)_x$ these defects are highly delocalized and mobile, so one can consider them as neutral solitons in a polymer chain. In the earliest studies of the electrical conductivity of polyacetylene it has already been assumed that charged spinless solitons play the role of charge carriers [14]. Kivelson [15] developed a theory of charge transport considering isoenergetic electron hopping between solitons localized near dopant ions. As the doping level increases at $y \approx 10^{-4}$ a decrease of the number of paramagnetic sites occurs, being accompanied by conductivity growth. At the same time, the contribution of charge hops to neutral solitons decreases, and it has been revealed that the conductivity in this regime is caused by the presence of spinless charge carriers. At dopant concentrations exceeding $y = 10^{-2}$ to 5×10^{-2}, a transition from isoenergetic electron hops to hops of variable range between localized sites is observed.

Peculiarities connected with the spin conservation law may be expected in the processes of electron (or hole) transport via paramagnetic sites $\dot{S}$. The experimentally detected sensitivity of the dark conductivity of polyacetylene to weak magnetic field [16] could be related to this very type of charge transport process.

To study the effect of a resonant microwave magnetic field on the electrical conductivity of polyacetylene a surface type sample with two Al electrodes was placed in a quartz ampoule into the cavity of the EPR spectrometer working in the X range. The sample has been doped within the ampoule by iodine vapor. The sample having resistance within the range 10^{4} to $10^{9}\,\Omega$ has been included in the electrical chain containing a d.c. voltage source, and load resistor $R_0 = 10^{5}\,\Omega$. The amplitude modulation of the microwave power has been used at 1300 Hz. The signal from the output resistor has been applied to a narrow band amplifier and lock-in detector and then to a recorder. The microwave field in the cavity was $B_1 = 1$ G. Under these conditions a change of the dark conductivity of the sample was revealed at the resonant magnetic field $B_0 = 3500$ G. Typical spectra registered are given for different samples in Fig. 6.4. The sign of the effect observed is positive: microwaves enhance the conductivity. The linewidth of the spectrum obtained exceeds by 8 to 10 G the linewidth of the ordinary EPR spectrum registered on the same samples by microwave power absorption.

The RYDMR spectrum detected by dark conductivity is the first example of detection of a spin-dependent process under the dark conditions.

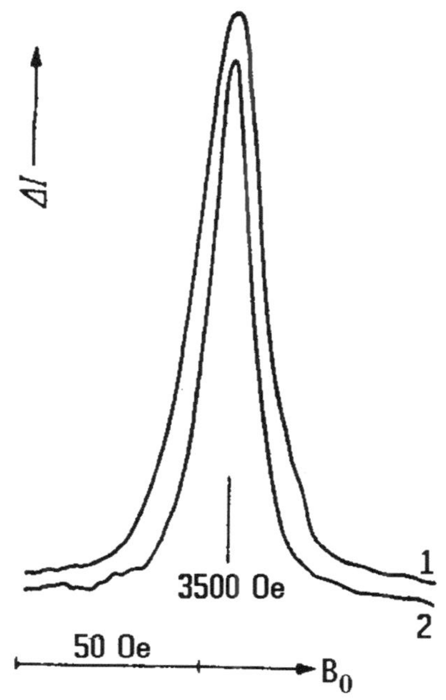

Figure 6.4 Spectra of the microwave power-induced resonant change of the dark current through the polycrystalline sample detected at room temperature: (1) undoped $(CH)_x$ having been exposed to air over one month. $R_{RT} \approx 3 \times 10^8\ \Omega$; (2) weakly iodine-doped *trans*-$(CH)_x$, $R_{RT} \approx 10^6\ \Omega$. From Frankevich et al. [10].

Nonequilibrium conditions in the population of the pair spin states are produced here due to the very process of conductivity in an electrical field.

A polaron–soliton pair responsible for the RYDMR spectra on the dark conductivity is shown schematically in Fig. 6.5. The rate of charge hopping inside the pair depends on the total spin of the pair. A hop of the charge inside the pair produces a pure singlet state: S^+ charged soliton plus a neutral site h^0. The detailed theory of the effect is given in Kubarev and Frankevich [18]. The same kind of effect was later found [19] on other materials all having a high concentration of free radicals as localization sites for charge carriers. The results obtained have shown that (1) in polyacetylene and similar materials neutral paramagnetic species take part in the carrier transport reacting with charge carriers with spin $S = 1/2\hbar$ and forming spinless charged sites; at the next step these sites may dissociate, and the pairs are reformed being shifted by the external electrical field. (2) Paramagnetic charged states, presumably polarons, take part in the conductivity of weakly doped polyacetylene; the increase in the level of doping leads to a decrease in the ratio of magnetosensitive to nonmagnetosensitive carrier transport processes.

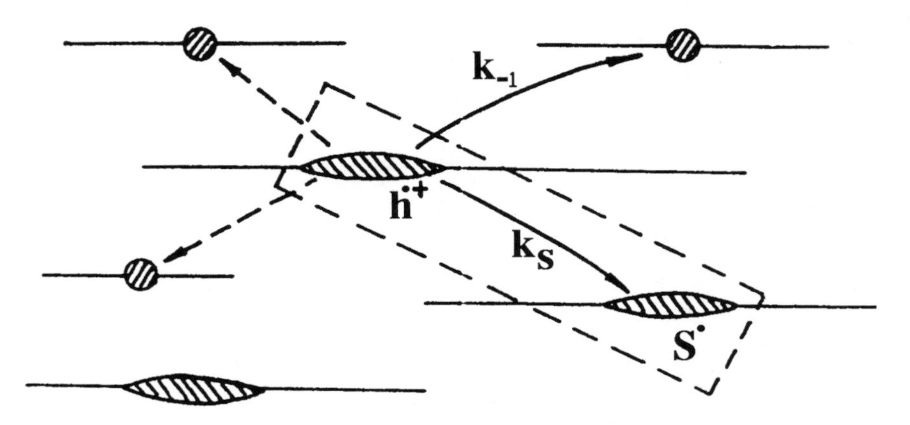

Figure 6.5 A polaron–soliton pair is shown inside the frame. The rate of charge hopping inside the pair depends on the total spin of the pair. A hop of the charge inside the pair produces a pure singlet state: S^+ charged soliton and neutral site h^0. From Frankevich et al. [17].

6.5. Conductivity and Spin Effects in Conducting Polymers

An important question concerns the nature of photoexcited states and photophysical processes in conducting polymers with degenerated and non-degenerated ground states. A photomodulation technique is usually used to study photogenerated states within the sample: neutral ($\dot{S}$) and charged ($S^{\pm}$) solitons, polarons p^-, and bipolarons $bp^{2\pm}$, triplet states. In this technique two beams are usually used: a pump beam (from an Ar^+ laser) for producing the excited states and a probe beam (from an incandescent light source dispersed by a monochromator) for measuring the changes in transmission. The photomodulation spectra, however, cannot resolve paramagnetic species from spinless ones. A combination of the photomodulation technique with magnetic resonant effect on the rate of the processes in pairs of paramagnetic species permits us to do that. Wei et al. [20] applied such a combined technique, which effectively is RYD—by an absorption—MR [but called by Wei et al. spin dependent photomodulation (SDPM)], to conducting polymers polyacetylene (PA) and poly(paraphenylenevinilene) (PPV). The technique was mentioned in Chapter 5 (see Fig. 5.13). Two types of spectra were taken: so called B_0-SDPM spectra where the wavelength λ of the probe beam was fixed and B_0 swept and P-SPDM spectra were B_0 was fixed and λ changed. It was revealed for PA that the absorption band in the region of $h\nu \approx 1.35\,\text{eV}$ belongs to neutral soliton $\dot{S}$ species and the band at $h\nu \approx 0.4\,\text{eV}$ to the charged but spinless ones $S^{\pm}$: the RYDMR signals obtained in both regions at 4 K was shown to be due to the common origin in a spin-dependent process of soliton annihilation

$$\dot{S} + \dot{S} \rightarrow {}^1(S^+ + S^-) \rightarrow \text{phonons} \rightarrow \text{ground state}$$

The microwave field increased the rate of the annihilation of $\dot{S}$, thus decreasing the light absorption at 1.35 eV and increasing that at 0.4 eV. Pairs of solitons are produced in a pure triplet state due to decay of the excited state of PA:

$$2A_g \rightarrow {}^3(S + S) + {}^3(S + S)$$

Thus the RYDMR technique has permitted resolution of the long-standing puzzle associated with the transient light-induced absorption spectra of PA.

In PPV the behavior of B_0-SDPM spectra at $h\nu = 0.35$ eV, $h\nu = 1.36$ and 2 eV has shown the different origin of the signals: the two first were due to the bipolarons, and the latter was due to optical transitions in the triplet manifold, associated with photogenerated triplet excitons in PPV. The latter does not decay in PPV into solitons. Polaron recombination is a spin-dependent process, the rate of which is increased under the action of microwave-induced transitions in polaron pairs. That leads to smaller bipolaron density as they are in equilibrium with polarons [20].

In conclusion, by the SDPM technique the following main photoexcitation processes in the degenerate (such as *trans*-PA) and nondegenerate (such as PPV) conducting polymers have been identified:

1. e–h pairs are photoexcited to a $1B_u$ state following above-gap photon absorption.
2. The $1B_u$ state decays into $S^{\pm}$ or $p^{\pm}$ or undergoes an intersystem crossing to the $2A_g$ state.
3. The $2A_g$ state decays into neutral solitons or triplets.
4. Neutral solitons recombine either to the ground state or to charged solitons; triplets recombine to the ground state or back into $2A_g$ and polarons either form bipolarons or recombine to the ground state.

6.6. RYDMR Spectra of Ion Radicals in Hydrocarbon Solids

Under the action of ionizing radiation on hydrocarbon solids geminate pairs of radical cations and electrons are produced. Essentially all of the subsequent chemistry follows from the spatial distribution of the cation–electron pairs and further transformation of the radical cations. Alkane radical cations (RH^+) are σ-radical cations, a fundamentally important class of organic intermediates. They occur in hydrocarbon radiolysis and their chemistry is prototypical of processes occurring in radiation modification of polymers and has considerable relevance to the understanding of the biological effects of ionizing radiation. In the past decade much has been learned about the geometry and electron structure of RH^+ species formed by positive charge transfer and stabilized in irradiated frozen matrixes such as $CFCl_3$ and other hydrocarbons using EPR spectroscopy (for review see [21]). Unfortunately, conventional cw EPR methods are unable to detect RH^+ species in neat alkanes even at 4 K. It is assumed that very rapid conversion of alkane radical cations into alkyl radicals occurs. Alkyl radicals are ubiquitous in irradiated alkanes and are easily detected by EPR. The advantage of RYDMR in studies of short-lived radical cations is due to its superior sensitivity and time domain capability relative to conventional EPR methods and to its high degree of spectral resolution (hyperfine structure) compared to condensed-phase optical experiments.

In the RYDMR method an intensity of fluorescence as a measure of the reaction rate in the pairs is usually being used. The latter arises when recombination between a radical cation and electron scavenged by an acceptor molecule A occurs [see a scheme (6.8)].

It was possible to observe the RYDMR spectra of cation radicals in saturated hydrocarbons at a temperature 35 K that were produced by the ionization of parent molecules [21]. That was established on the basis of their hyperfine structure. With anthracene-d_{10} used as acceptor-scintillator A the only radical ions that can give rise to a hyperfine structure in the RYDMR spectrum are radical cations. Werst and Trifunac [21] call the spectra obtained as fluorescence detected magnetic resonance (FDMR) spectra. To obtain the spectra in [21] an electron beam pulse (12 ns long) from a 3-MeV accelerator started the reaction. The time position of pulses that govern the microwaves and the detection of the fluorescence is shown in Fig. 5.15. An increase of the temperature up to the melting of the sample led to conversion of original radicals, and spectra of only allyl radicals and their dimers could be obtained. These results have permitted us to arrive at the conclusion that ion-molecule reactions play an essential role in the transformation of ion radicals, however, they do not work at low temperature. Recombination in pairs in solid matrices, which gives rise to excitation of scintillator molecules A, appears to occur as a result of electron tunneling between A^- and the cation radical.

RYDMR spectra of the pair "cation radical–solvated electron" in solid glassy solutions of *p*-terphenyl in squalane irradiated by X-rays were observed at 173 K [22]. The spectra were monitored by different parameters, namely by fluorescence of *p*-terphenyl (the spectrum had a positive sign corresponding to recombination of singlet pairs) and by phosphorescence of *p*-terphenyl (the spectrum was the same but negative).

6.7. RYDMR of Short-Lived Pairs of Triplet Excitons

6.7.1. Resonant Transitions in a Two-Triplet Pair

The RYDMR method has bright prospects for studying pairs of triplet species and, above all, triplet excitons in molecular solids and solutions. The measurement of variations in the exciton annihilation rate due to resonant transitions between the Zeeman levels of a pair formed in the reaction of the type depicted in Eq. 6.9 is basic to the RYDMR method. The structure of the energy levels of a two-triplet pair in a magnetic field is shown in Fig. 6.6. The energy levels are obtained by summing up the energies of the corresponding triplet levels

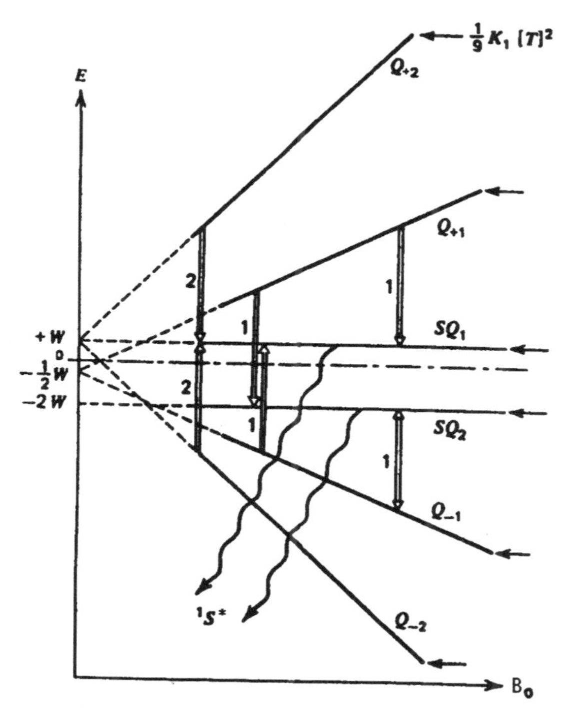

Figure 6.6 Energy levels of a pair of equivalent noninteracting triplet excitons.

constituting the levels of a pair, on the assumption that zero-field splitting between the triplet and singlet as well as between the quintuplet and singlet component are zero. In a strong external magnetic field $B_0 \gg D, E$ (where D and E are the fine splitting parameters in triplet excitons) for a pair of equivalent triplet excitons, two of nine spin states of the pair contain a singlet component with an amplitude $|S_i|^2$ equal to 1/3 and 2/3. These states are as follows:

$$SQ_1 = |00\rangle = (1/\sqrt{3})|S\rangle + \sqrt{2/3}|S\rangle \qquad (6.16)$$

and

$$SQ_2 = (1/\sqrt{2}))|- \ +\rangle + |+ \ -\rangle) = (1/\sqrt{3})|Q_0\rangle - \sqrt{2/3}|S\rangle \qquad (6.17)$$

The singlet components are mixed with quintuplet components. The rest of the states of a two-triplet pair are purely triplet and quintuplet ones, Q_{+1} and $Q_{\pm 2}$. On exposure to a resonant microwave field, transitions that may involve a change in the projection of the magnetic quantum number m by $\Delta m = \pm 1$, with the total spin S preserved ($\Delta S = 0$), are possible. Hence, of all conceivable transitions, only those that involve the levels containing a singlet component are essential. The triplet states of a pair of equivalent triplet excitons do not contain any singlet admixture, so transitions between them would not affect the total population of the levels that do contain a singlet. But if a pair of nonequivalent triplets is formed or a configuration interaction occurs between them, the singlet component will be present in states of any multiplicity. It is worth noting that the population of the quintuplet levels Q_{+1} and Q_{-1} having no singlet components are much higher than those of mixed level SQ_1 and SQ_2. This is due to the fact that the population rates of every level are equal ($\frac{1}{9}k_1 n_T^2$, where n_T is the triplet exciton concentration), whereas the lifetimes of singlet-containing states $\tau_{SQi} = (k_{-1} + k_S|S_i|^2)^{-1}$ are short compared to those of purely quintuplet states, $\tau_Q = k_{-1}^{-1}$. As a consequence, an applied microwave field mainly induces transitions from quintuplet to singlet levels, thereby building up their population. A total change of the population of the levels SQ_1 and SQ_2 can be expressed as follows:

$$\Delta n = \frac{1}{9} k_1 \frac{k_S k_C(B_0)}{k} |S_1|^2 |S_2|^2 \left(\frac{1}{k_{-1} + k_S|S_1|^2} + \frac{1}{k_{-1} + k_S|S_2|^2} \right) \tag{6.18}$$

Here $k_C(B_0)$ is the rate constant of resonant transitions due to a microwave field:

$$k_C(B_0) = B_1^2 \left(\frac{g\beta}{\hbar} \right)^2 \frac{\tau}{1 + (g\beta/h)^2 (B_0 - B_{res})^2 \tau^2} \tag{6.19}$$

where B_1 is the amplitude of a microwave magnetic field in the cavity, and $|S_2|^2 = 1 - |S_1|^2$. A change of the total rate constant $\gamma = \Sigma \gamma_i$ of triplet exciton annihilation due to resonant transitions is calculated from the value Δn and gives

$$\frac{\Delta\gamma}{\gamma} = k(B_0)|S_1|^2|S_2|^2 \frac{k_S R}{k_{-1}(k_{-1} + k_S/2)} \tag{6.20}$$

where R is weakly dependent on $|S_1|^2$ and varies from 1 to 1.25 for values of $|S_1|^2$ from 0 to 1 (at $k_S \approx k_{-1}$). For typical values of the parameters $\tau = 2 \times 10^9$ s^{-1} and $|S_1|^2 = 0.5$, $\Delta\gamma/\gamma$ was found to be

$$\Delta\gamma/\gamma \approx 10^{-5} B_1^2 \quad \text{or} \quad \Delta\gamma/\gamma \approx (\omega_1 \tau)^2 \tag{6.21}$$

where $\omega_1 = g\beta B_1/\hbar$.

Thus at $B_1 = 1$ G the rate constant of triplet exciton annihilation is expected to change by a factor of about 10^{-5}. This value is many orders of magnitude larger than variations in microwave power across a detector brought about by resonant absorption, when use is made of conventional EPR spectrometry.

6.7.2. First Experiments with RYDMR

Typical experiments on triplet pairs involve delayed fluorescence due to triplet exciton annihilation [see reaction (6.9) and the process that is the reverse of the annihilation, namely, quenching of fast fluorescence during the fission of singlet excitons into two-triplet pairs]. In the two cases the variations of the fluorescence intensity $\Delta L/L \sim \Delta\gamma/\gamma$.

The early experiments concerned magnetoresonant effects on the intermediate two-triplet complex [23] studied by fast fluorescence at room temperature of a 0.3-cm^2 tetracene single crystal placed in the cavity of the EPR optical detection spectrometer. An alternate sample was a polycrystalline powder of an anthracene–dimethylpyromellitimide charge-transfer complex. Figure 6.7 shows the spectrum of variations in tetracene fluorescence intensity for crystal orientation such that the *ab* plane was parallel to B_1 and B_0, whereas the *b*-axis formed an angle of about 35° with B_0. When the crystal was rotated in the *ab* plane, the positions of the spectral lines changed, just as in the case of an EPR spectrum of triplet excitons; the signal phase corresponded to a fluorescence intensity reduced under the microwave power effect. Figure 6.8 shows a similar spectrum taken for an anthracene–dimethylpyromellitimide (A-DMPT) polycrystalline charge-transfer complex. Transitions with $\Delta m = \pm 1$ and $\Delta m = \pm 2$ were observed. The shape of this spectrum is typical of an EPR spectrum for randomly oriented triplet excitons. The fluorescence intensity here was found to increase under the microwave power effect. One can directly define the lifetime of a two-triplet pair from its spectral linewidth: for tetracene $\tau = 5 \times 10^{-9}$ s ($\delta B_0 = 21$ G); for A-DMPT $\tau = 3.7 \times 10^{-9}$ s ($\delta B_0 = 15$ G).

Knowing the spectra line splitting as a function of the A-DMPT crystal orientation in a magnetic field, one can establish the fine structure parameters of the triplet excitons. In the case of polycrystals of A-DMPT these parameters are derived from the position of the spectral tails and peaks represented in Fig. 6.8. They proved to be $|D| = 750$ G and $|E| = 87$ G. These values of D and E are close to those for triplet excitons of anthracene and indicate that delayed

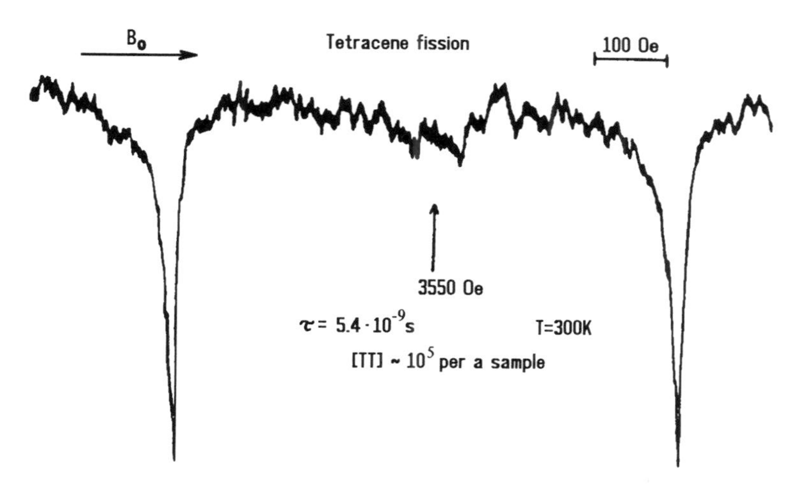

Figure 6.7 RYDMR spectrum of triplet exciton pairs in tetracene detected by prompt fluorescence. From Frankevich et al. [23].

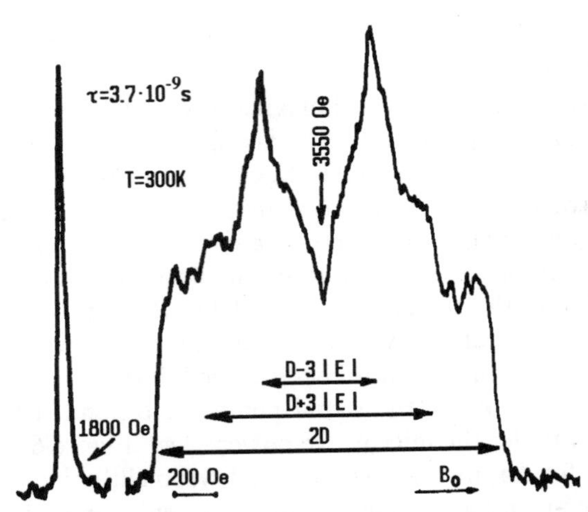

Figure 6.8 RYDMR spectrum of polycrystalline sample of an A-DMPT charge-transfer complex detected by delayed fluorescence. From Frankevich et al. [23].

fluorescence of charge-transfer complexes is the result of the annihilation of donor triplets.

A sharp dip in the center of the RYDMR spectrum of a polycrystalline sample distinguishes it from a conventional EPR spectrum of triplet excitons. At this point magnetic resonance occurs for triplet excitons oriented at a "magic" angle at which the levels SQ_1 and SQ_2 are degenerate. In this case only one of nine spin states contains a singlet component; hence $|S_1|^2 = 1$ and, according to formula (6.18), $\Delta\gamma/\gamma = 0$. This feature is closely studied by Lesin et al. [24] on an A-TCNB charge-transfer single-crystal complex. These crystals exhibited delayed fluorescence caused by the annihilation of triplet excitons localized on an anthracene molecule. The RYDMR spectrum of triplet excitons generally consists of two lines with positive polarities. The crystallographic axes a, b, and c of the A-TCNB single crystal coincide with those of the fine-splitting tensor. When a single crystal is rotated in a plane parallel to the vector of a spectrometer static magnetic field, the distance between two lines of the RYDMR spectrum is expressed by

$$\Delta H_{12} = 3[D(\cos^2\theta - \tfrac{1}{3}) + E(\cos^2\alpha - \cos^2\beta)] \tag{6.22}$$

where α, β, and θ are the angles formed by the a-, b-, and c-axis with the vector B_0. When the angle θ varied from 60 to 85°, the lines ran together and their intensity did not increase, as would have been the case with a conventional EPR spectrum of triplets, but, on the contrary, reduced to zero. These results are represented in Fig. 6.9. Figure 6.10 is the predicted RYDMR spectrum of a triplet pair in the crossing region of levels SQ_1 and SQ_2. Evidently the sign of the effect is expected to reverse in crossing, to the same as in radical pairs. This is the same spin-locking effect discussed in Sections 5.2 and 5.3 for radical pairs.

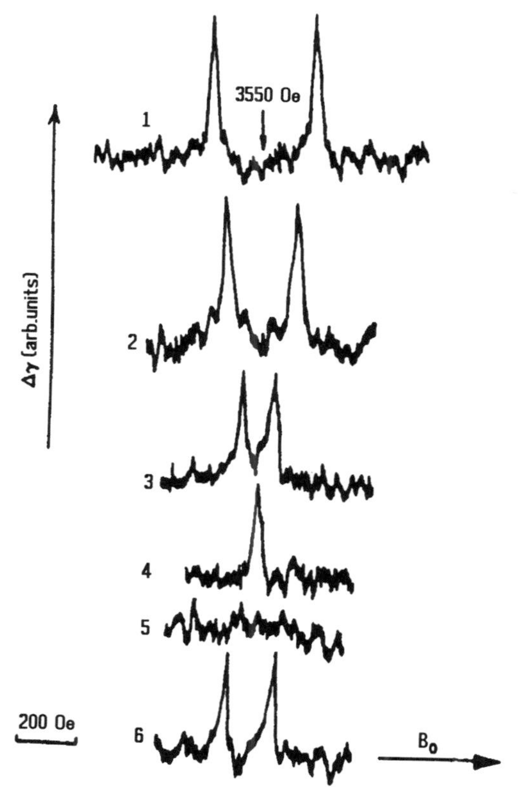

Figure 6.9 Variations in the RYDMR spectrum of a triplet exciton pair in a single crystal of an anthracene–tetracyanbenzene charge-transfer complex arising on crystal turning in the magnetic field of a spectrometer B_0. Spectra 1 to 6 correspond to angles $\theta = 60$, 68, 71, 75, and 83°, respectively. From Lesin et al. [24].

6.7.3. Spin Locking Effect for Triplets

The shape of RYDMR spectral lines for a pair of triplet excitons in crystalline hydrocarbons was calculated for the first time by Lesin et al. [24]. These calculations made use of the Johnson–Merrifield equation [6] for the spin-density matrix ρ of a T–T intermediate complex of two triplet excitons in a crystal, with due reference to exciton interaction with a high-frequency magnetic field B_1. The initial equation for density matrix ρ is as follows:

$$\frac{\partial \rho}{\partial t} = -\frac{i}{\hbar}(\hat{H}_0, \rho) - \frac{1}{2} k_S(P^{(S)}\rho + \rho P^{(S)}) - k_{-1}\rho + \gamma \tag{6.23}$$

where $\hat{H}_0$ is the total spin Hamiltonian of an exciton pair. This Hamiltonian can be written as

$$\hat{H}_0 = \hbar\omega_0(S_1^z + S_2^z) + H_1 + H_2 + V_{12} \tag{6.24}$$

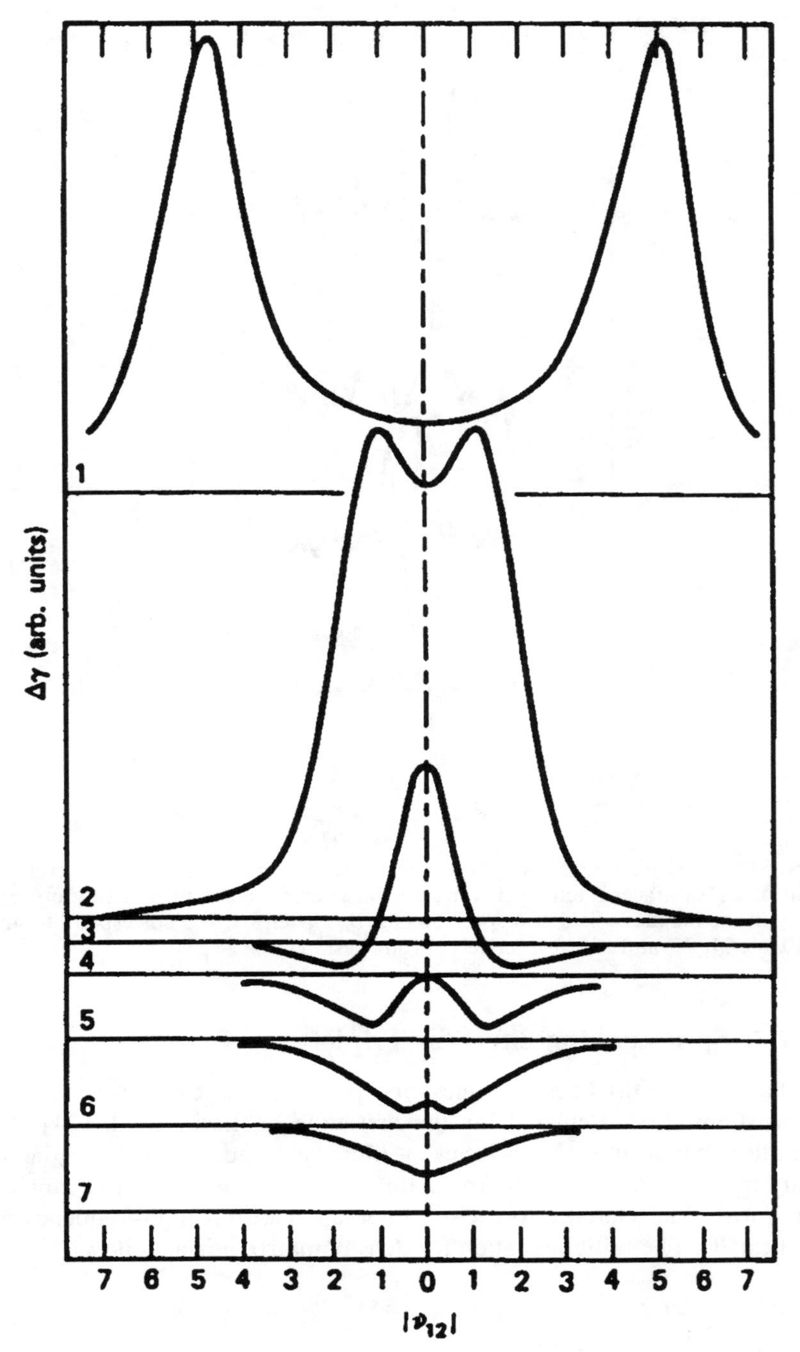

Figure 6.10 Calculated variations in the RYDMR spectrum of a triplet exciton pair arising from a changing energy difference ΔE between the levels SQ_1 and SQ_2. Curves 1 to 7 correspond to the values of ΔE measured in the linewidths 10, 3.3, 1, 0.7, 0.5, 0.25, and 0 respectively. The abscissa is a deviation of the magnetic field from a resonant one, and it is measured in linewidths. From Lesin et al. [24].

where H_1 and H_2 are the spin Hamiltonians of the zero-field splittings of the first and second excitons of the pair, $\hbar\omega_0 = g\beta B_0$, and V_{12} is the exciton interaction (of electrostatic or magnetic nature):

$$V(t) = \hat{V}[\exp(i\omega t) + \exp(-i\omega t)] \tag{6.25}$$

$$\hat{V} = \hbar\omega_1(S_1^x + S_2^x), \qquad \hbar\omega_1 = g\beta B_1$$

where B_1 is the induction of an electromagnetic field and ω is its frequency. The parameters k_{-1} and k_S have the sense of the processes of a fission process of the pair due to exciton separation and triplet–triplet annihilation via the singlet channel, respectively. In Eq. (6.23) $P^{(S)}$ is the projection operator onto the singlet state of the pair, and γ is the source operator.

Equation (6.23) was solved assuming that the probability for microwave field-induced transitions between energy levels is very low compared to the constants k_{-1} and k_S. This made it possible to use the approximation method over degrees of small parameter ω_1. The variation $\Delta\gamma(\omega)$ in the rate constant of triplet–triplet annihilation due to an a.c. magnetic field was calculated with an accuracy of ω_1^2, that is, in an approximation that is linear with respect to the microwave power.

The dependence $\Delta\gamma(\omega)$ was explicitly established for two limiting cases: (1) when the splitting between levels of the T–T pair that contain the S component is very large compared to the width of the levels and (2) over the range of values $k_S \ll k_{-1}$.

In the former case $\Delta\gamma(\omega) > 0$ for all ω and the RYDMR spectrum consists of two individual Lorentzian lines. In the latter case the behavior of $\Delta\gamma(\omega)$ is more complicated. If the splitting greatly exceeds the value of k_{-1}, the shape of the spectrum is just as in case (1); but as the value k_{-1} increases, spectral lines tend to broaden and run together, the sign of the effect is changed, and $\Delta\gamma(\omega)$ becomes negative.

6.7.4. Annihilation of Prepolarized Triplets

A RYDMR spectrum displays some interesting features if prepolarized triplet excitons take part in the formation of two-triplet pair. The intensities and directions of magnetoresonant transitions between Zeeman levels of an intermediate pair of triplet excitons are determined by the populations of these levels. It has been thought so far that each of the nine levels of a complex has the same probability of being populated at a rate $\frac{1}{9}k_1 n_T^2$, where k_1 is the rate constant of triplet exciton collision and n_T is the triplet exciton concentration. In this case the difference between level populations necessary for magnetoresonant transitions to be observed arose owing to the fact that the populations of the singlet-containing levels were consumed to form the singlet products of the reaction. The populations of these levels were pumped by magnetoresonant transitions that manifest themselves in a more intense delayed fluorescence (the sign of the RYDMR spectral lines was positive).

As is known, triplet excitons in molecular crystals may be polarized, that is, their Zeeman sublevels $|0\rangle$, $|+\rangle$, and $|-\rangle$ may have unequal populations. When all the levels of triplet excitons are equally populated, all the levels of a two-triplet pair are being filled at the same rate. However with polarized exciton spins the rates of population the levels of the pair will be different (see Fig. 6.6). In the former case resonant transitions induced by a microwave field will contribute to populations of the central, singlet-containing levels, since their populations are lower than those of the quintuplet levels, being consumed in the reaction channel. In the case of polarized exciton spin there may occur a sign reversal of the variations in the delayed fluorescence intensity, provided that singlet-containing levels prove to be more populated, on account of polarization, than quintuplet levels. Such a sign reversal of the spectral line was observed experimentally by Frankevich et al. [25] for a certain A-TCNB crystal orientation. This led to the conclusion that in triplet excitons the populations of the levels $|\pm 1\rangle$ and $|0\rangle$ are not equal. For crystals with prepolarized triplet excitons the absorption of microwave power by free excitons that have not yet formed a pair is also manifest in the RYDMR signal. Excitons whose polarizations has not reversed owing to resonant transitions would impart this change to the polarizations of pairs yet to be formed. Hence a pair's lifetime measured by its linewidth will be equal to the time of spin–lattice relaxation. For example, in the RYDMR spectra of A-TCNB crystals linewidths are found to be 0.8 G, which corresponds to a pair lifetime $T_1 = 1.4 \times 10^{-7}$ s.

6.7.5. Double Resonant Transitions in Two-Triplet Pairs

If the lifetime of triplet pairs is rather long, it is quite probable that at a sufficiently high microwave power the triplet pair may undergo double transitions. To these correspond transitions from the levels $|++\rangle$ (in a weak field) and $|--\rangle$ (in a strong field B_0) to the level $|0\,0\rangle$ containing a singlet component (Fig. 6.11). To observe such transitions by means of delayed fluorescence the following three conditions should be met: (1) a high microwave power that saturates single transitions should be used, (2) pairs should have long lifetimes ($\tau_{pair} \approx 10^{-7}$ s), and (3) the levels $|++\rangle$ and $|0\,+\rangle$ (or $|--\rangle$ and $|0\,-\rangle$) should have different populations. The latter condition requires prepolarized triplet exciton spins. Figure 6.12 shows the RYDMR spectra for two crystal orientations at various microwave power levels. It has revealed that for an "abnormal" negative polarity a narrow line of the opposite polarity appears against the background of a broad line of the spectrum. A reduction of the microwave power would decrease the intensity of the narrow line and eventually make it disappear. Such spectral behavior corresponds to expectations for double transitions in two-triplet pairs [25].

Similar spectra were obtained in A-TCNB crystals for pairs of unpolarized triplet excitons formed during the detection of resonant transitions in a zero magnetic field at a temperature of 1.2 K [26]. As was shown by von Schütz et

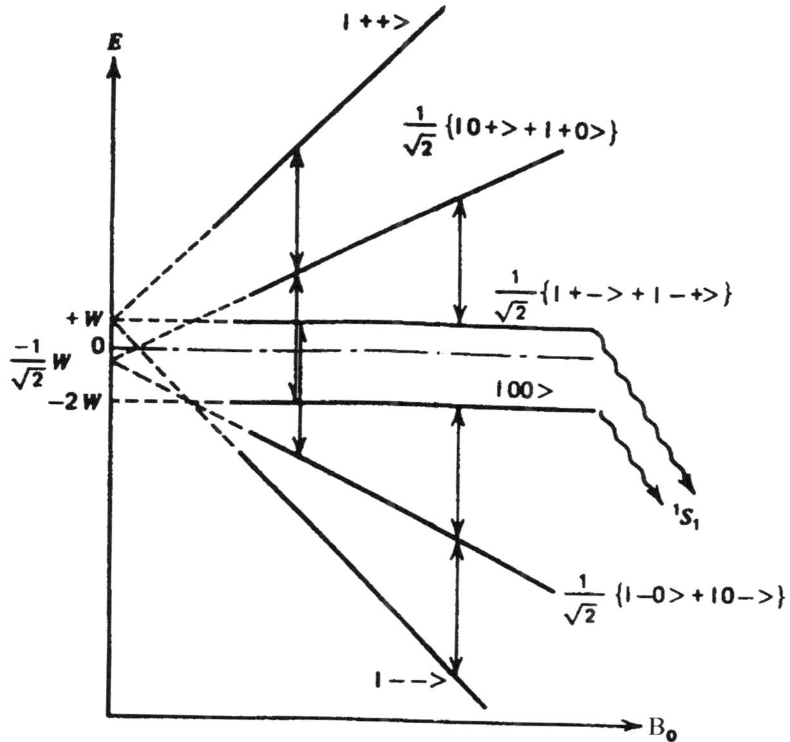

Figure 6.11 Diagram of Zeeman levels of a two-triplet pair in a strong magnetic field B_0, with double transitions mixing the states $|++\rangle$ and $|--\rangle$ with the singlet containing state $|00\rangle$.

al. [27] the sign reversal of the line in this case is related to a nonmonotonic dependence of fluorescence intensity on the rate constant of triplet exciton annihilation. This effect was attributed to the phosphoroscope used to separate delayed fluorescence from exciting light. The rate constant of triplet annihilation increased monotonically with microwave power.

6.7.6. Magnetic Resonance of Intermediate Triplet–Doublet Pairs

Interaction of triplet excitons with doublet species is observed in essential reactions in photochemistry and photobiology, for it gives rise to two effects: quenching of a triplet excited states and energy transfer from a triplet to a doublet. The reaction scheme was presented earlier by (6.12). A RYDMR spectrum of the pair ($^3D\cdots{}^2R$) was experimentally detected by Frankevich et al. [28] by variations ΔL in the intensity L of delayed fluorescence due to triplet exciton annihilation, provided that the lifetime of triplet excitons was

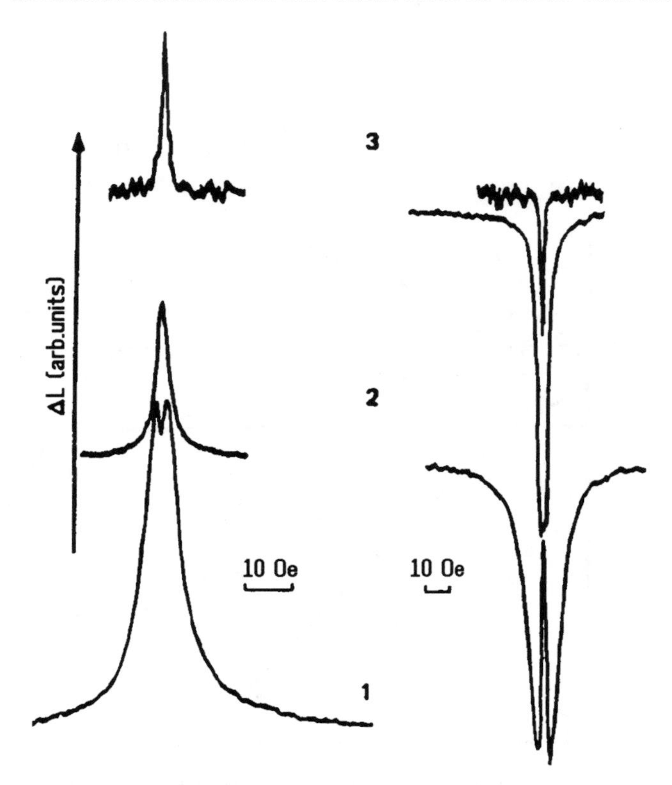

Figure 6.12 Shape of the RYDMR spectral line of a two-triplet pair in a crystal of A-TCNB charge-transfer complexes for different microwave powers. The ordinate shows variations in delayed fluorescence intensity due to resonant transitions. (1) Full microwave power (~ 3 W); (2) attenuation = 10 dB; (3) attenuation = 20 dB. In the measurements the vector $\mathbf{B}_0$ was in the *ab* plane of the crystal; for the spectra on the left $\mathbf{B}_0 \parallel \mathbf{b}$ ($\alpha = 0$); for the spectra on the right $\alpha = -45°$. From Frankevich et al. [25].

determined by their interaction with free radicals. A-DMPT charge-transfer crystals were used. In these crystals triplet excitons are localized on anthracene molecules and migrate in the *bc* planes of the crystals. The lifetime of triplet excitons is 1.8 ms. RYDMR spectra of delayed fluorescence for the initial polycrystalline A-DMPT samples were compared with those of polycrystalline A-DMPT samples containing free radicals produced therein (Fig. 6.13). The bottom spectrum differs from the top one by a single line, with a g-factor of 2 superimposed on the somewhat less intense initial spectrum. This line arises from the spectrum of radicals in pairs (^{3}D$\cdots^2$R). The number of radicals in the sample was estimated from the extent of delayed fluorescence quenching to be $\sim 10^{11}$. The total scheme of the transitions occurring in the pair (^{3}D$\cdots^2$R) is given in Fig. 6.14, where one can see transitions in triplets and doublets manifesting themselves by a changed yield of the reaction via doublet channel. It is interesting to note that when one uses the reaction of triplet quenching by

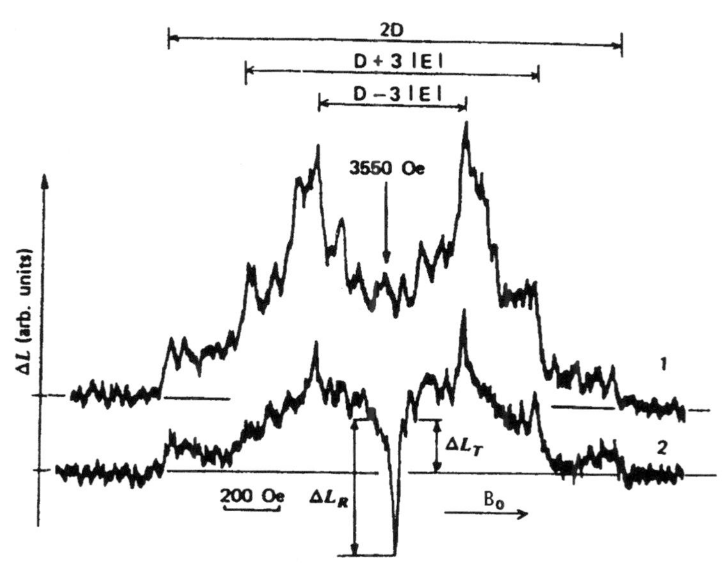

Figure 6.13 RYDMR spectra of short-lived pairs of paramagnetic species in a polycrystalline sample of A-DMPT charge-transfer complex. ΔL is the variation in intensity of delayed fluorescence of a sample. (1) Nonirradiated sample: the spectrum corresponds to a pair of triplet excitons of anthracene. (2) The same sample irradiated by the dose of 0.3 Mrad. The central line in the spectrum is due to microwave power absorption by radicals in triplet exciton–radical pair. From Frankevich et al.

radicals as well as RYDMR spectrum detection, one succeeds in obtaining a signal from a number of radicals (10^{11}) that is inaccessible to the most sensitive direct EPR methods. The value of the variation of the rate constant Q [see Eq. (6.14)] in a resonant microwave field was calculated by Frankevich et al. [28] based on the kinetic approach previously developed [23]. Within the framework of this approach, an intermediate pair is treated as a particle having a system of energy levels, each characterized by its own lifetime and fission time constant. Here the microwave power effect on the reaction rate constant k_l [see the reaction scheme (6.12)], which is a function of the steady-state populations of pair levels, reveals itself as a variation in these populations. The kinetic scheme gives a qualitatively good picture of a spectrum of resonant variations in the rate constants away from crossing of intermediate pair levels. Assuming that the values of the rate constants for transitions between levels $i \rightarrow j$ due to a microwave field B_1 are far less than that of the constant k_{-1}, one finds ΔQ as a function of B_1, B_0, k_{-1}, and k_1. The spectrum obtained has three lines of positive polarity: one from radicals, and two of equal intensity from triplet excitons. In the calculations made no account was taken of the contribution of spin–lattice relaxation to the spectral linewidth, since the pairs under consideration had lifetimes less than the spin relaxation time of the system. This assumption is justified by the observation of the RYDMR spectrum of these

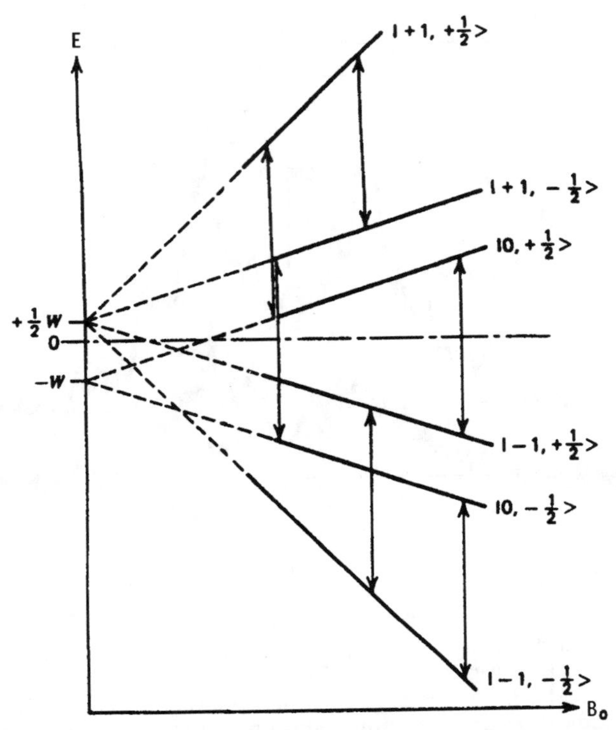

Figure 6.14 Diagram of the energy levels of a triplet exciton–radical pair in a strong magnetic field. The energy W is determined by a fine splitting in the triplet excitons and depends on the crystal orientation with respect to the constant magnetic field B_0. The wave function of each level and the resonant transitions in triplet excitons and radicals that affect the rate of reaction are indicated.

pairs as well as by RYDMR lines that are much broader (equal to 20 G in our case) than those of the typical EPR spectra (1 to 2 G) of these particles.

An interesting application of the RYDMR method to investigation of surface paramagnetic centers was made by Pristupa et al. [29]. It consisted of the creation of spin correlated pairs produced by the triplet molecular excitons interacting with the surface centers. Two cases were studied. In the first a polycrystalline layer of polyacene dye was deposited on the fresh surface of the AgBr crystal. In the second such a layer was applied to the ruby crystal. The dye was excited by the light, and RYDMR spectrum may be seen by monitoring the delayed fluorescence. The spectrum observed was the result of pairs of nonoriented triplet excitons but in the center of the spectrum a line with g-factor $g = 2$ and the width of 40 G has appeared (for AgBr). In the case of ruby crystal the position of the central line depended within the range of 400 G on the orientation of the crystal in the external magnetic field. In the second case the line appeared only under external pressure. Pristupa et al. [29] believe that the paramagnetic centers on the surface of AgBr are $Ag(^2S_{1/2})$

atoms; on the ruby crystal they are Cr^{3+} ions. Pristupa and Sakun [30] claim that these centers are involved in the evolution of spin in the pairs of triplets so the evolution actually proceeds not in pairs but in triads. However, the possibility is not excluded that triplet excitons are being quenched by surface centers producing pairs "triplet exciton-center," and then survived triplets produce T–T pairs that still maintain information about the quenching. The technique used permits us to "feel" on the surface about 10^{10} to 10^{12} center/cm^2.

References

1. Pope, M., Swenberg, Ch. E. *Electronic Processes in Organic Crystals*, Vol. 1 and 2. Clarendon Press, Oxford, 1982.
2. Noyes, R. M. *J. Chem. Phys.* **1954**, *22*, 1349–1360.
3. Onsager, L. *Phys. Rev.* **1938**, *54*, 554–557.
4. Johnson, R. C., Merrifield, R. E., Avakian, P., Flippen, R. B. *Phys. Rev. Lett.* **1967**, *19*, 285–289.
5. Merrifield, R. E. *J. Chem. Phys.* **1968**, *48*, 4318–1926.
6. Johnson, R. C., Merrifield, R. E. *Phys. Rev.* **1970**, *B1*, 816–828.
7. Altweg, L. *Chem. Phys. Lett.* **1979**, *63*, 97–101.
8. Geacintov, N. E., Pope, M., Fogel, F. *Phys. Rev. Lett.* **1969**, *92*, 593–598.
9. Ern, V., Merrifield, R. E. *Phys. Rev. Lett.* **1968**, *21*, 609–614.
10. Frankevich, E. L., Kadyrov, D. I., Sokolik, I. A., Pristupa, A. I., Kobryanskii, V. M., Zurabyan, N. J. *Phys. Stat. Sol. (b)* **1985**, *132*, 283–295.
11. Frankevich, E. L., Pristupa, A. I., Tribel, M. M., Sokolik, I. A. *Dokl. Akad. Nauk SSSR* **1977**, *236*, 1173–1179.
12. Frankevich, E. L., Tribel, M. M., Sokolik, I. A., Pristupa, A. I. *Phys. Stat. Sol (b)* **1978**, *87*, 373–385.
13. Su, W. P., Schrieffer, J. R., Heeger, A. J. *Phys. Rev.* **1980**, *B22*, 2099–2214.
14. Heeger, A. J., McDiarmid, A. J. *Mol. Cryst. Liquid Cryst.* **1981**, *77*, 1–15.
15. Kivelson, S. *Phys. Rev.* **1982**, *B24*, 3798–3809.
16. Frankevich, E. L., Sokolik, I. A., Kadyrov, D. I., Kobryanskii, V. M. *Zh. Eksper. Teor. Fiz., Pisma* **1982**, *36*, 401–405. English translation: *JETP Lett.* **1982**, *36*, 486–489.
17. Frankevich, E. L., Kadyrov, D. I., Sokolik, I. A., Kobryanskii, V. M., Pristupa, A. I., Zurabian, N. J. *Mat. Sci.* **1984**, *10*, 73–76.
18. Kubarev, S. I., Frankevich, E. L. *Soviet J. Chem. Phys.* **1990**, *3*, 1502–1600. Original in Russian: *Khim. Fiz.* **1984**, *3*, 964–972.
19. Kadyrov, D. I., Minasyan, G. G., Sokolik, I. A., Poperechenykh, V. I., Frankevich, E. L. *Soviet J. Chem. Phys.* **1990**, *6*, 2641–2648. Original in Russian: *Khim. Fiz.* **1987**, *6*, 1343–1350.
20. Wei, X., Hess, B. C., Vardeny, Z. V. *Phys. Rev. Lett.* **1992**, *68*, 666–670.
21. Werst, D. W., Trifunac, A. D. *J. Phys. Chem.* **1991**, *95*, 3466–3477.

22. Saik, V. O., Anisimov, O. A., Molin, Yu. N. *Chem. Phys. Lett.* **1985**, *116*, 138–144.

23. Frankevich, E. L., Pristupa, A. I., Lesin, V. I. *Chem. Phys. Lett.* **1977**, *47*, 304–309.

24. Lesin, V. I., Sakun, V. P., Pristupa, A. I., Frankevich, E. L. *Phys. Stat. Sol. (b)* **1977**, *84*, 513–522.

25. Frankevich, E. L., Pristupa, A. I., Lesin, V. I. *Chem. Phys. Lett.* **1978**, *54*, 99–105.

26. Stende, W., von Schütz, J. U. *J. Lumin.* **1979**, *18/19*, 191–200.

27. von Schütz, J. U., Stende, W., Wolf, H. C., Yakhof, V. *Chem. Phys.* **1980**, *46*, 53–64.

28. Frankevich, E. L., Lesin, V. I., Pristupa, A. I. *Zh. Eksper. Teor. Fiz.* **1978**, *75*, 415–422. English translation: *Soviet Phys. JETP* **1978**, *48*, 208–214.

29. Pristupa, A. I., Lesin, V. I., Krasotkina, I. A. *Chem. Phys. Lett.* **1991**, *180*, 569–575.

30. Pristupa, A. I., Sakun, V. P. *Khim. Fis.* **1992**, *11*, 1118.

CHAPTER 7

RYDMR in Liquid Solutions

7.1. RYDMR in Radical Ion Pairs Generated by Ionizing Radiation

The action of penetrating radiation on materials is one of the very important processes studied by photochemistry, radiation chemistry, and radiobiology.

The primary particles acting in the substance are electrons, irrespective of the nature of primary radiation (γ-rays, X-rays, or fast electrons). An electron ejected from a molecule in the condensed phase in the process of ionization suffers many collisions and loses its initial kinetic energy. If the energy becomes as low as thermal energy when the electron is still not too far from the positive ion, within a distance shorter than the Onsager radius r_{ons}, then one can say that an ion–electron pair is produced. There is a high probability that such a pair will recombine geminately. The Onsager radius is determined as the distance at which the energy of the Coulomb interaction between ion and electron equals the thermal energy kT. In hydrocarbons at room temperature $r_{\text{ons}} \approx 200$ Å. Recombination inside the pair proceeds within a short time, which may be estimated as a drift time of an electron to the ion through the distance $a \approx r_{\text{ons}}$ under the action of the Coulomb field:

$$\tau_{\text{pair}} = \frac{1}{3}\frac{\varepsilon a^3}{\mu e} \approx \frac{6 \times 10^{-12}}{\mu} \tag{7.1}$$

Here μ is the sum of mobilities of electron and ion, in $cm^2/V\,s$; ε is the dielectric constant; at numerical estimation $a \approx 100$ Å and $\varepsilon = 3$ were taken.

At typical mobility of a quasi-free electron in hydrocarbon liquids at room temperature $\mu = 1\ \mathrm{cm^2/V\,s}$; the lifetimes of the pair prove to be about $\tau_{pair} \approx 10^{-11}$ s. Taking into account the diffusion at distances of about r_{ons} gives the same estimation. Lifetimes estimated are not long enough for spin evolution, and methods based on the magnetic field effect on the spin state of the pair are not operative. However, if acceptor molecules are present in the solution in high enough concentration, the electrons have a good chance of being scavenged, and the mobility of charged particles μ is reduced significantly. In most cases the lifetime of the pair of ion radicals proves to be adequate for spin evolution and mixing of spin states.

The primary part of the transformations induced by ionizing radiation in liquid hydrocarbons goes through the stage of the ion–electron pair production. Radical cations and their geminate electron partners are thought to be the primary chemical entities produced by ionizing radiation. Essentially all of the subsequent chemistry follows from the spatial distribution of cation–electron pairs and further transformation of the radical cations and electrons. The production and recombination of ion-radical pairs in a solution containing solvent molecules M and solute molecules D and A are described by the following reactions:

$$\begin{aligned}
&\text{(a)}\ \mathrm{M} \rightarrow \mathrm{M}^+ + e^- \qquad\qquad (7.2)\\
&\text{(b)}\ e^- + \mathrm{A} \rightarrow \mathrm{A}^-\\
&\text{(c)}\ \mathrm{M}^+ + \mathrm{D} \rightarrow \mathrm{D}^+ + \mathrm{M}\\
&\left.\begin{aligned}
&\text{(d)}\ {}^1(\mathrm{M}^+ + e^-, \mathrm{A}^-) \rightarrow {}^1\mathrm{M}^* \text{ or } {}^1\mathrm{A}^*\\
&\text{(e)}\ {}^1(\mathrm{D}^+ + e^-, \mathrm{A}^-) \rightarrow {}^1\mathrm{D}^* + \mathrm{A} \text{ or } \mathrm{D} + {}^1\mathrm{A}^*
\end{aligned}\right\}\ \text{geminate recombination}
\end{aligned}$$

7.1.1. Experimental Investigations of Reactions in Nonpolar Solutions

An important feature of a nonpolar medium is the geminate recombination of ions arising from transfer of a positive charge from a solvent ion M^+ to an impurity D and electron capture by A. One can imagine also other reactions involving initial ions M^+:

Ion–molecular charge transfer among solvent moleules or impurity molecules

$\mathrm{M}^+ + \mathrm{M} \rightarrow \mathrm{M} + \mathrm{M}^+$, $\quad \mathrm{A}^- + \mathrm{A} \rightarrow \mathrm{A} + \mathrm{A}^-$.

Decay of primary ions $\quad \mathrm{M}^+ \rightarrow$ smaller ions.

Ion–molecular reactions leading to rearranged, protonated, dimerized, etc. ions $\quad \mathrm{M}^+ + \mathrm{M} \rightarrow$ new ions.

Ion–molecular reactions in nonpolar solutions are of particular interest, however, they cannot be studied by the standard EPR technique because of the very low concentration of radical ions produced under radiolysis due to their short lifetime.

As far as the RYDMR method permits us to obtain the EPR spectrum during the lifetime of ions in pairs, it is possible, in principle, to study such processes experimentally. Anisimov et al. [1] were the first to detect a magnetic resonance spectrum for the ionic pairs ($C_{10}H_8^+ \cdots C_{10}H_8^-$), produced by irradiation of 10^{-2} *M* solution of naphthalene in squalane. Naphthalene was used as an acceptor for both electrons and holes. The ionizing radiation consisted of fast positrons emitted by radioactive ^{22}Na. Since the emission of positrons is accompanied by the escape of a γ-quantum, it was possible to synchronize the measurement of fluorescence with the ionizing pulse. By introducing a delay one could measure the fluorescence of rather long-lived pairs of ion radicals with lifetime over 10^{-9} s. When a sample in the cavity under the magnetic field of a magnetic resonance spectrometer was exposed to a microwave field of fixed frequency, the photon count rate was reduced according to the resonant conditions of microwave power absorption. With biphenyl molecules admixed to squalane, magnetic resonance spectra have been obtained for biphenyl ions with a resolved hyperfine structure [2].

Trifunac and Smith [3] produced ionization in solutions of pyrene in decalin by periodic 5- to 55-ns pulses of fast electrons from a Van de Graaf accelerator. The microwave pulse was 100 ns long. In addition, one could change the duration of the microwave pulse with respect to the ionizing pulse as well as introduce a controlled delay between the microwave pulse and the instant at which detection of recombination fluorescence began. A more comprehensive investigation carried out by the same authors [4], who applied the method to a system of 10^{-3} *M* diphenyl oxazole in cyclohexane, has demonstrated the feasibility of deriving RYDMR spectra for ion radical pairs with lifetimes ranging from 30 ns to 4 μs, thereby detecting a decrease of the fluorescence from the products of pair geminate recombination. These results, in particular, have shown that in nonpolar liquid solutions in the time interval that is essential for the formation of the RYDMR signal (10^{-9} to 10^{-7} s), the lifetime distribution function of ion radicals is well approximated by the power function $F(t) \sim t^{-3/2}$.

In a number of works of Anisimiv and co-workers [5–9] the nature of radicals and many processes in pairs produced under radiolysis were studied. They have used X-rays for the irradiation (see Section 5.5) and monitored the products of reactions in pairs by recombinational fluorescence. They called the modification of RYDMR method used optically detected electron paramagnetic resonance (OD EPR). The RYDMR method permits us to identify radicals that take part in recombination by their hyperfine structure. An example is the identification of radicals by the RYDMR spectrum of a solution containing 10^{-3} *M* perfluoronaphthalene and 10^{-3} *M* durole in liquid squalane [10]. The hyperfine structure of the spectrum has shown the pairs to consist of cation radicals of durole and anion radicals of perfluoronaphthalene. In some cases the RYDMR method proves to be adequate for studying free charges such as electrons and holes. The latter term is often used for positive ions of the solvent molecules that are able to migrate by charge transfer. As noted above, the geminate recombination of electron and hole in liquid hydrocarbons at room temperature takes only $\sim 10^{-12}$ s and it is too fast to permit the evolution of the spin. However, by lowering the temperature it is possible to make the recombination slower and thus to provide conditions for observation of RYDMR spectra of quasi-free electrons. Anisimov et al. [11] were able to observe the RYDMR spectra for the liquid solution of durole (10^{-2} *M*) in squalane at $T < 250$ K. A line was seen the intensity of which increased as the temperature was lowered, but decreased when electron acceptors were added. A decrease in the mobility of electrons was achieved also by the solvatation of electrons by water added to the same system [11]. That allowed us to observe the spectrum of the hydrated electron even at room temperature.

The RYDMR spectrum of the hole is difficult to observe because of a fast reaction of charge transfer to donor molecules of admixtures. When the latter are absent one can expect the spectrum of the mobile hole not to show any hyperfine structure, as during the movement of the hole by charge transfer between solvent molecules the hyperfine interaction will be averaged to zero. Such an RYDMR spectrum was observed on a benzene solution containing 10^{-4} *M* *para*-terphenyl during irradiation by X-rays [10]. A part of these spectra is shown in Fig. 7.1. Molecules of *para*-terphenyl added played the role of electron acceptors A and simultaneously that of donors D and of scintillators. The fluorescence appeared in the process (7.2e), however, the spin of ion radical D^+ at the moment of its recombination was conditioned by its evolution as the spin of the hole while it was in a state M^+ that was a precursor of D^+ [see the process (7.2c)]. It is why the RYDMR spectrum at low enough concentration of *para*-terphenyl corresponded to that of M^+, that is, of the hole (Fig. 7.1b). At a higher concentration of *para*-terphenyhl the time of stay of the charge and spin in the state M^+ was diminished below the value necessary for spin evolution in the magnetic field, and the RYDMR spectrum converted to the spectrum of *p*-terphenyl$^+$ ions (Fig. 7.1a). The signal produced by the hole became lower when electron donor molecules of *p*-xylene were added. The signal of ion radicals of xylene with well-known hyperfine structure appeared (Fig. 7.1c). At still a higher concentration of *p*-xylene when charge transfer between ions and molecules of *p*-xylene became possible, a narrow structureless signal appeared.

The RYDMR method enables estimation of the rate constant k_D of the charge-transfer process by using the width of the RYDMR line of the hole, which is determined by the fast exchange:

$$\Delta\omega = \sqrt{3/2}\Delta^2/(k_D c) \tag{7.3}$$

Here $\Delta\omega$ is the width of the spectrum (in frequency units), Δ^2 is the second moment of the spectra of cation radical in the absence of charge transfer, and c is the concentration of molecules. Values of k_D are of the order of 10^8 to 10^9 M^{-1} s^{-1} for dodecane, pentadecane, *p*-xylene, and toluene.

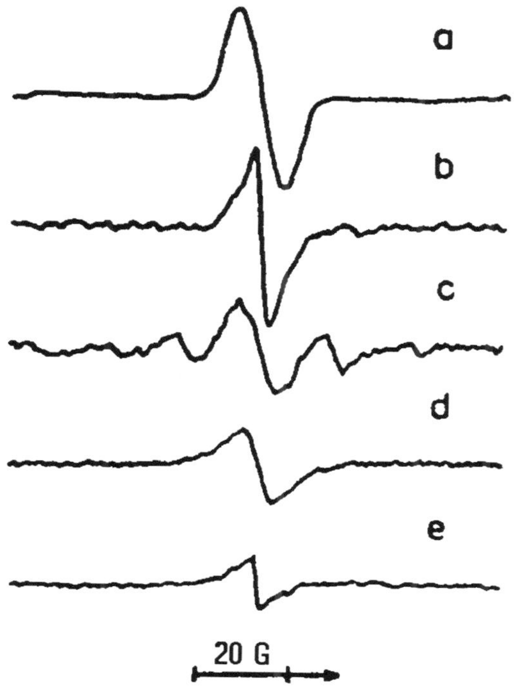

Figure 7.1 Variations of RYDMR (or OD EPR) spectra of liquid solutions under irradiation by X-rays. (a, b, c, d) Spectra of benzene solutions with impurities: (a) 10^{-3} *M* *p*-terphenyl, (b) 10^{-4} *M* *p*-terphenyl, (c) 10^{-4} *M* *p*-terphenyl + 10^{-2} *M* *p*-xylene, (d) 10^{-4} *M* *p*-terphenyl + 1 *M* *p*-xylene; (e) 10^{-4} *M* *p*-terphenyl in *p*-xylene. From Molin et al. [10].

The capture of holes by electron-donor molecules gives rise to the appearance of ions D^+. Collision of the latter with neutral molecules D leads to charge transfer. The collisions also produce cation radicals of dimers D_2^{+}, which reveal themselves in the RYDMR spectra [9].

An influence of ion–molecule charge transfer on the RYDMR spectra of nonpolar liquid solution has been studied [8]. A theoretical analysis has shown that ion–molecule charge transfer increases the width of the spectral lines in the region of slow exchange when $A\tau_e \gg 1$, where A is the hyperfine interaction constant and τ_e is the time of charge transfer. A fast exchange is expected to make the lines narrower. The dependence of the width of the spectra (between points of the maximum slope) on the concentration of acceptor molecules enables us to determine the rate constant of charge transfer in the region of fast exchange. The rate constants of charge transfer between electron acceptor molecules and molecular anions were measured in the process of geminate recombination of the latter with cations in the following radical–ion pairs: hexafluorobenzene$^-$/deuteroanthracene$^+$, biphenyl$^-$/biphenyl$^+$, naphthalene$^-$/naphthalene$^+$. The radical pairs were generated in squalane and isooc-

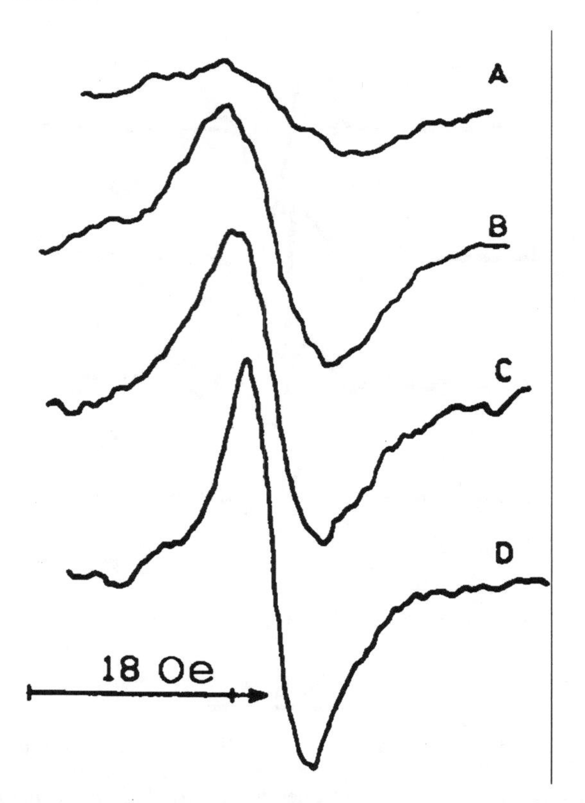

Figure 7.2 First-derivative biphenyl$^-$/biphenyl$^+$ RYDMR (or OD ESR) spectra in isooctane at 21°C. Concentration of biphenyl increases from A to D: 10^{-3}, 5×10^{-3}, 10^{-2}, and 2×10^{-2} *M*, respectively.

tane solutions containing corresponding acceptors of electrons and holes under X-ray irradiation. Figure 7.2, taken from Saik et al. [8], shows as an example concentration-transformed RYDMR spectra for biphenyl$^-$/biphenyl$^+$ in isooctane at the biphenyl concentrations corresponding to the spectral narrowing. The results have shown that in the isooctane solvent the anion–molecular charge transfer process is principally induced by diffusional collisions of reacting particles, and rate constants are close to those calculated as $k_D = 8RT/(30\eta)$, where R is the gas constant and η is the solvent viscosity in centipoise ($k_D = 1.3 \times 10^{10}\ M^{-1}\,s^{-1}$ at room temperature).

To study primary processes of radiolysis a pulse technique of the RYDMR (see Section 5.5.2.) has been used [3, 4, 12]. The technique permitted us to ionize the solution by a fast electron beam during 5 to 15 ns, then to switch on the microwave power and to open the time window for measuring the fluorescence. This technique is called fluorescence detected magnetic resonance (FDMR). Changing the time delay between pulses of ionization and microwave action permitted us to reveal ion radicals in pairs that were created during the lifetime of the pair from primary particles M^+ and e^-. The technique was applied [12] for the investigation of primary processes of radiolysis of liquid-saturated hydrocarbons. RYDMR spectra that have been carried out in neat alkanes can be divided into liquid phase and solid state experiments. The principal goal was to characterize the initial, solvent radical cations by direct real time detection of their EPR spectra using pulse RYDMR technique. Methods previously used to investigate radicals cations in alkanes, such as conductivity and optical spectroscopy, do not provide a unique structural signature for radical cations as does EPR, and thus identification of radical cations by these methods is tenuous. Conventional EPR does not possess sufficient sensitivity or time resolution to detect radical cations in neat alkanes. The hyperfine structure of the RYDMR spectra of radiolized alkanes has allowed identification of the ion radicals observed as the products of the decay of primary ions into olefine ion and H_2 molecule. Primary ions themselves were not observable even at zero delay time between ionization and microwave pulses. It followed from this result that their lifetime was shorter than 10 ns. However, in solid solutions ion radicals of saturated hydrocarbons were revealed (see Section 6.6) as lowering the temperature makes the decay of primary ion radicals slower.

When impurity molecules with the ionization potential lower than that of the solvent are present in the saturated hydrocarbon solution then the production of cation radicals of impurity molecules takes place. The RYDMR method permits us to obtain their magnetic resonance spectra and to study their structure in a manner similar to what is being done by the technique of matrix isolation but in conditions of free motion of ions. This technique was applied [13, 14] to study the structure of ion radicals *cis*-decalin$^+$, tetramethylbutane$^+$, dicyctopentyl$^+$, and others.

A change of the RYDMR spectra in the dependence on the time delay between ionization and microwave pulses is evidence of the production of aggregate of ions of the type X_n^+, where $n \geqslant 2$. It was seen in methylcyclohexane containing anthracene as a scintillator and tetramethylethylene as X, and also in cyclohexane with pyrene [12].

Production of dimeric cation radicals proceeds in the reaction of secondary ions D^+ with neutral molecules of impurities D. The processes of production of ions of the type D_2^+ have been studied [9]. Ions D_2^+ (molecules of *p*-terphenyl were impurities D and *trans*-decalin was a solvent) produced the RYDMR spectra with hyperfine splitting, which was two times less compared with that of original monomer cations. An increase of the concentration of impurity D led to broadening of the lines and to a shift of the components of the spectrum to its center. A theory was developed by Lukzen et al. [15] that permits us to calculate rate constants for a dimerization process starting from experimental saturation parameters of the RYDMR spectra.

Ionizing radiation produces ion radicals in liquid solutions in the singlet state of pairs. However, such pairs may also be produced by laser light as a result of multiphoton processes, and some triplet states may be involved as intermediate ones. The sign of the RYDMR spectrum permits to clarify the mechanism of ion production: a negative sign indicates the singlet precursor of the pair [16].

7.1.2. Experimental Investigations of Reactions in Polar Solutions

Certain features exist in studying the radiolysis of liquid solutions such as alcohols. First of all the lifetime of primary radicals is very short due to the existence of fast ion–molecule reactions of the type

$$ROH^{\cdot +} + ROH \rightarrow RO^{\cdot} + ROH_2^+ \tag{7.4}$$

that convert ion radicals into neutral radicals. Second, the high dielectric constant of liquid alcohols compared with that of saturated hydrocarbons makes the Onsager radius r_{ons} shorter, and a smaller part of the ions recombines geminately, and it takes a shorter time. The probability that a radical ion will escape its geminate partner is given by Eq. (7.5), where r_{ons} is the Onsager radius [remember that it is the geminate pair separation distance at which the Coulomb energy equals the thermal one, i.e., $e^2/(4\pi\varepsilon_0\varepsilon r_{ons}) = kT$] and a is the initial separation between positive and negative ions:

$$p = \exp(-r_{ons}/a) \tag{7.5}$$

In hydrocarbons where $\varepsilon \approx 2$, the Onsager radius is on the order of 300 Å and the fraction of ions that escape is exceedingly small (<5%). In alcohols where ε may range as high as 35 or greater, the Onsager radius is about 20 Å (methanol ≈ 16 Å, ethanol ≈ 19 Å, 2-propanol ≈ 30 Å), and the fraction of escaped ions can easily exceed 50%.

The geminate ion lifetime is strongly affected by the Onsager radius and for polar liquids it is expected to be more than two orders of magnitude smaller than for hydrocarbons. The ion mobilities are affected by both the temperature and the viscosity of the liquid, decreasing as the temperature is decreased and the viscosity is increased. As a consequence, the geminate ion lifetime can be extended by lowering the temperature and geminate ion pairs can be more easily detected at low temperature.

RYDMR spectra of ion radicals of impurities in alcohols and ethers have been obtained [17]. By using an impurity PPO (2,5-diphenyl-1,3-okazol) as a scintillator Percy et al. [17] were able to obtain the spectra produced by two types of pairs, namely, $PPO^+ \cdots PPO^-$ and $e_S^- \cdots PPO^+$. Solvated electron e_S^- was identified by its narrow line ($\Delta B_0 = 5$ G), which appeared in the spectra when the temperature was lowered to 203 K. At room temperature the spectrum of solvated electrons was observed at the photoionization of a solution of tetramethylphenylendiamin (TMPD), and also of perylene in 2-propanol [16] by the light of an eximer laser at $\lambda = 308$ nm. The hyperfine structure of cation radicals was observed.

Use of radiowaves of lowest possible frequencies has certain advantages when polar solutions are studied. It minimizes losses of the power connected with the interaction of radiowaves with electrical dipoles. A low-field EPR spectrometer at a frequency of 100 MHz was used [18] to study reactions in pairs of ion radicals generated by X-ray irradiation of polar solutions. Resonant transitions for particles with g-factors of 2 had a place at $B_0 = 36$ G. The sample was placed inside an induction coil of the LC resonant circuit. RYDMR spectra were observed for aromatic electron acceptors in tetrahydrofurane, ethanol, acetone, acetonitryl, and other polar solvents under irradiation. High amplitudes of B_1 could be achieved easily and that permitted us to invert narrow lines in the RYDMR spectrum by using the spin-locking effect. That method helps identify the narrow signal of the solvated electron on the background of a broad spectrum of ion radical pairs. When the values of the field B_0 and B_1 are of the same order of magnitude, some nonlinear effects appear. They reveal themselves producing multiquanta transitions between magnetic sublevels and giving peaks in RYDMR spectra at the B_0 field values $B_0 = n\omega_0\hbar/g\beta$, where $n = 2, 3, 4$, and ω_0 is the main working frequency of the spectrometer. The intensity of peaks caused by multiquanta transitions increases with B_1 as B_1^n. A simple theory of that phenomenon was suggested by Abraham [19], and its application to RYDMR spectra was developed by Morozov et al. [20].

7.2. RYDMR in Pairs of Neutral Radicals

Evolution of spin of pairs in liquids proceeds on the distances between partners, which are much larger than intermolecular distances. The probability of the repeated contacts for neutral particles is much less than unity. The Coulomb attraction of charged particles increased this probability many times. No one was able to see RYDMR spectra of neutral radicals in homogeneous liquid solutions. However, the technique of isolation of radical pairs within the micelles prevents a fast dissociation of pairs and enables us to investigate the interaction and chemical reactions of the radicals by the magnetic field effects using permanent field B_0 and microwave field B_1.

The work of Turro and Cherry [21] was the first in which a large influence of the isolation of pairs in micelles was observed. Practically all the works devoted to magnetic field-sensitive reactions in micelles deal with photoexcitation, dissociation, and transformation of carbonyl compounds (ketones). As substances producing micelles they used mainly sodium dodecyl sulfate (SDS), sodium octyl sulfate (SOS), and sodium decyl sulfate (SDeS). Radical pairs are produced in the following processes:

1. Breaking of the α-bond in ketone molecules. A typical substance is dibenzylketone (DBK) $PhCH_2COCH_2Ph$:

$$\mathrm{DBK} + h\nu \rightarrow {}^1(\mathrm{DBK})^* \rightarrow {}^3(\mathrm{DBK}) \tag{7.5}$$

$$ {}^3(\mathrm{DBK}) \rightarrow {}^3(\mathrm{PhCH_2\dot{C}O\cdots\dot{C}H_2Ph}) \rightleftharpoons {}^1(\mathrm{PhCh_2\dot{C}\cdots\dot{C}H_2Ph})$$

2. Photoreduction of quinones in the process of H atom transfer from the donor molecule DH or from the micelle-producing compound to a triplet excited molecule. A typical substance is benzophenone (BP) $Ph_2C{=}O$:

$$\mathrm{BP}+h\nu \longrightarrow {}^1(\mathrm{BP})^* \longrightarrow {}^3(\mathrm{BP}) \xrightarrow{+\mathrm{DH}} {}^3(\mathrm{Ph_2\dot{C}OH}\cdots\dot{\mathrm{D}}) \rightleftharpoons {}^1(\mathrm{Ph_2\dot{C}OH}\cdots\dot{\mathrm{D}}) \quad (7.6)$$

After leaving the micelle and being in the free volume the ketyl radicals $Ph_2\dot{C}OH$ can dimerize.

3. Carbonyl compounds in the excited state can act as electron acceptors producing pairs of ion radicals with oxidized donor molecules.

The micelles usually are present in a water–alcohol medium, their concentration being about 10^{-4} to 10^{-3} M. The shape of micelles is spherical with a radius of about 18 Å; a single micelle consists of from 30 to 60 molecules of SDS. The internal volume of the micelle contains a homogeneous nonpolar medium. A typical time of the diffusion of radicals "to the wall" is about 200 ns. The detailed measurements of parameters of micelles are possible by studying the kinetics of the decay of fluorescence of the pyrene solution within the micelle excited by a laser pulse [22]. A small internal size of a micelle where radicals can diffuse makes us believe that the prominent part of their lifetime will be spent suffering on exchange interaction. An averaged exchange interaction can reveal itself in RYDMR spectra.

Reactions in pairs of free radicals produced by photoreduction of menadione or anthraquinone in micelles SDS have been studied [23, 24]. Okazaki et al. [23, 24] investigated the influence of the resonant microwave power on the yield of free radicals in a water solution where radicals became scavenged by spin traps producing stable nitroxide radicals. Their concentration was accumulated during the time of irradiation of micellar solution by light (during 60 s) and was measured in the same cavity where a microwave field influenced the reaction. Okazaki et al. called the technique used as product yield detected electron spin resonance (PYESR) and remarked that under the action of a resonant microwave field it was possible to vary the yield of products up to 20%. A selective action of microwaves in different lines of the RYDMR spectrum splitted by hyperfine interaction was demonstrated. It enables us to influence the yield of product molecules containing certain isotope atoms. (For more details on the magnetic isotope effect see Chapter 9.)

Reactions in pairs of radicals in SDS micelles in a solution of toluene or benzene initiated by photoexcitation of benzophenone were studied by the RYDMR method [25]. Radicals were produced under the action of pulsed light of an eximer laser ($\lambda = 308$ nm, 20 ns, 100 mJ, 10 Hz) as a result of H atom addition from donor cyclohexane. McLauchlan and Nattrass [25] measured the concentration of diphenylketyl radicals in micelles by the intensity of their fluorescence excited by a second pulse of light with $\lambda = 337$ nm and a duration of 6 ns. The microwave field B_1 switched on also as a pulse of 150 ns duration with controlled delay time after photoexcitation pulse (see Section 5.5.2). A hyperfine structure of cyclohexadienyl radical was revealed in the RYDMR spectra. An increase of B_1 up to 13 G did not lead to reversing of the spectra although some evidence of the existence of exchange interaction in pairs with $J \simeq 4$ G were revealed by the static magnetic field effect.

References

1. Anisimov, O. A., Grigoryants, V. M., Molchanov, V. K., Molin, Yu. N. *Chem. Phys. Lett.* **1979**, *66*, 265–268.
2. Anisimov, O. A., Grigoryants, V. M., Molin, Yu. N. *Zh. Eksper. Teor. Fiz., Pisma* **1979**, *30*, 589–693. English translation: *JETP Lett.* **1979**, *30*, 555–558.
3. Trifunac, A. D., Smith, J. P. *Chem. Phys. Lett.* **1980**, *73*, 94–98.
4. Smith, J. P., Trifunac, A. D. *J. Phys. Chem.* **1981**, *85*, 1645–1653.
5. Molin, Yu. N., Anisimov, O. A., Grigoryants, V. M., Molchanov, V. K., Salikhov, K. M. *J. Phys. Chem.* **1980**, *84*, 1853–1856.
6. Anisimov, O. A., Grigoryants, V. M., Molin, Yu. N. *Chem. Phys. Lett.* **1980**, *74*, 15–18.
7. Molin, Yu. N., Anisimov, O. A. *Radiat. Phys. Chem.* **1983**, *21*, 77–82.
8. Saik, V. O., Luksen, N. N., Grigoryants, V. M., Anisimov, O. A., Doktorov, A. B., Molin, Yu. N. *Chem. Phys.* **1984**, *84*, 421–430.
9. Saik, V. O., Anisimov, O. A., Lozovoy, V. V., Molin, Yu. N. *Z. Naturforsch.* **1985**, *40a*, 239–245.
10. Molin, Yu. N., Anisimov, O. A., Melekhov, V. I., Smirnov, S. N. *Faraday Disc. Chem. Soc.* **1984**, *78*, 289–301.
11. Anisimov, O. A., Molin, Yu. N., Smirnov, S. N., Rogov, V. A. *Radiat. Phys. Chem.* **1984**, *23*, 727–729.
12. Werst, D. W., Trifunac, A. D. *J. Phys. Chem.* **1991**, *95*, 3466–3477.
13. Werst, D. W., Trifunac, A. D. *J. Phys. Chem.* **1988**, *92*, 1093–1097.
14. Werst, D. W., Bakker, M. G., Trifunac, A. D. *J. Am. Chem. Soc.* **1990**, *112*, 40–46.
15. Lukzen, N. N., Saik, V. O., Anisimov, O. A., Molin, Yu. N. *Chem. Phys. Lett.* **1985**, *118*, 125–129.
16. Bakker, M. G., Trifunac, A. D. *J. Phys. Chem.* **1991**, *95*, 550–556.
17. Percy, L. T., Werst, D. W., Trifunac, A. D. *Radiat. Phys. Chem.* **1988**, *32*, 209–215.
18. Koptyug, A. V., Saik, V. O., Anisimov, O. A., Doktorov, A. B., Morosov, S. V., Antzutkin, O. N. 24th International Congress on Magnetic Resonance and Related Phenomena, AMPERE. Poland, Posnan, 1988.
19. Abraham, A. *Principles of Nuclear Magnetism.* Clarendon Press, Oxford, 1961.
20. Morozov, V. A., Antzutkin, O. N., Koptyug, A. V., Doktorov, A. B. *Mol. Phys.* **1991**, *73*, 517–525.
21. Turro, N. J., Cherry, W. R. *J. Am. Chem. Soc.* **1978**, *100*, 7431–7435.

22. Atik, S., Nam, M., Singer, L. *Chem. Phys. Lett.* **1979**, *67*, 75–78.

23. Okazaki, M., Sakata, S., Konaka, R., Shiga, T. *J. Phys. Chem.* **1987**, *86*, 6792–6798.

24. Okazaki, M., Shiga, T., Sakata, S., Konaka, R., Toriyama, K. *J. Phys. Chem.* **1988**, *92*, 1402–1407.

25. McLauchlan, K. A., Nattrass, S. R. *Mol. Phys.* **1988**, *65*, 1483–1503.

CHAPTER

8

RYDMR In Photosynthetic Systems

8.1. Charge Separation in Primary Photosynthetic Processes

8.1.1. Introduction

The conversion of the energy of light into the energy of chemical bonds of the synthesized chemical compounds proceeds in processes of photosynthesis. The energy of light is used in photosynthesis with a high efficiency, and thus studying the mechanism of reactions involved is essential both for the proper understanding of the photosynthesis itself in living systems—in plants and bacteria—and for the simulation of similar processes in biomimetic systems. The processes occurring during the main steps of photosynthesis just after light absorption are of primary importance. The light is being absorbed in the systems by pigments, most important of which are chlorophyll molecules. In cells of the photosynthesizing systems chlorophyll molecules are distributed nonuniformly and are concentrated mainly in functional photosynthetic units that enter in larger formations called chloroplasts [1, 2]. Each cell contains from 20 to 100 chloroplasts. Within a photosynthetic unit a large number of light-harvesting molecules in antenna works on the photochemically active molecule of chlorophyll that forms together with the primary donor and acceptor, and also with some other molecules, the so-called reaction center. Electronic excitation from antenna is transferred in a resonance process to reaction centers where the main photochemical process, the separation of charges, takes place.

It is well known that the primary photochemical reaction is a reversible oxidation–reduction transformation of chlorophyll in the reaction center. In the next steps an electron moves by charge transfer in the acceptor system and eventually is used to reduce the molecule of nicotinamide-adenine-dinucleatide-phosphate (NADPH), which reacts in dark reactions of synthesis of hydrocarbons from carbon dioxide. The positive charge from the reaction center is used for oxidation of water molecules. Though the organization of the transport and utilization of electrons are different for higher plants and for photosynthesizing bacteria, the primary processes of the charge separation are similar in these living systems. They are being studied usually on the separated reaction centers of photosynthesizing bacteria, which may then be transferred to a liquid solution. Reaction centers may even be converted to crystalline form [3].

8.1.2. Reactions in Reactions Centers

Chlorophyll molecules in a reaction center of bacteria (bacteriochlorophyll, BChl) are usually in the form of dimers $BChl_2$ and they work as a primary donor of electrons. The primary acceptor molecule A_1 is bacteriopheophytine (BPh), and the secondary acceptor A_2 is the molecule of quinone, Q. Dissociation of primary pairs $BChl_2^+ \cdots BPh^-$ consists in the transfer of the electron from BPh^- to Q and takes about 200 ps. This time is too short to permit an essential effect on the spin state of the pair. Therefore to study primary pairs it is necessary to reduce the molecules of quinone, Q, artificially or to remove them from reaction centers. It gives pairs $BChl_2^+ \cdots BPh^-$ no way to dissociate but leaves it possible for ion radicals to recombine inside the pair.

At the recombination in the singlet channel dimers $BChl_2$ are produced both in the ground and in an excited state. Triplet channel gives triplet excited dimers $^3(BChl_2)$. Production of triplet dimers reveals itself in the transient spectrum of optical absorption of reaction centers. Induced triplet–triplet absorption of the probing light serves as a measure of the reaction rate in primary triplet pairs. The processes described are shown in Fig. 8.1. The absorption of the probing light by chlorophyll molecules in the ground state may be used also to measure the reaction rate in pairs. The population of the ground state reflects the transformation of a part of the dimers into triplets and ions, and the part may be large enough at high intensity of excitation, which is usually provided by laser pulses. The intensity of fluorescence of excited chlorophyll molecules formed by recombination in pairs may be used to measure the reaction rate in the singlet channel.

The lifetime of primary pairs that are blocked in respect to dissociation is determined by the rate of recombination only and was found to be 10 to 20 ns. It enables them to be investigated by methods based on magnetic spin effects. It is worth noting that no other methods possess the sensitivity and specificity adequate to detect the processes in primary pairs. Secondary pairs of the type $BChl_2^+ \cdots BPh \cdots Q^-$ are more long living (up to 100 μa), and it is possible to apply an ordinary EPR technique for their study.

The detailed mechanism of the charge transfer from the excited donor

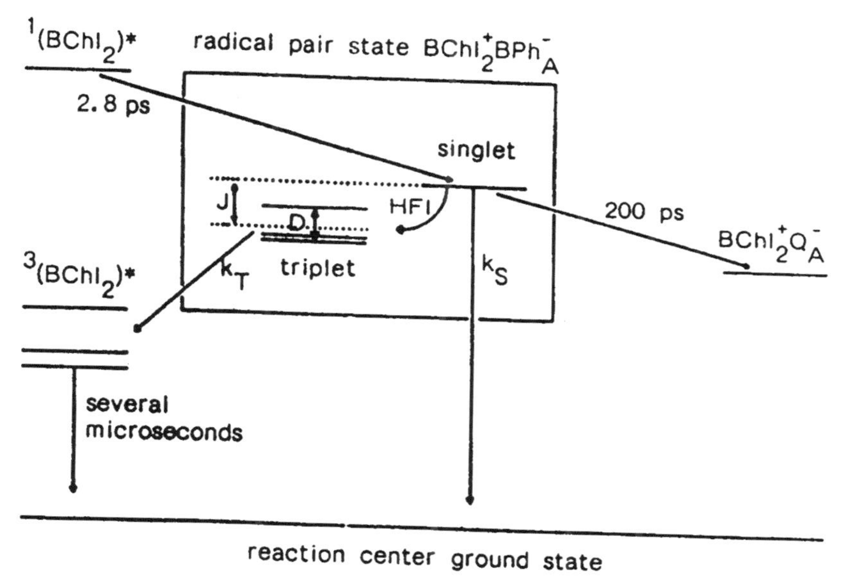

Figure 8.1 Schematic presentation of the electron transfer step in bacterial reaction centers. The energy level spacings in the figure are not true to scale. From Lersch et al. [12].

$^1(BChl_2)^*$ and production of the pair $BChl_2^{+} \cdots BPh^{-}$ is not yet quite clear. It is possible that an electron in the first step is transferred onto an intermediate acceptor $BChl_A$, which is situated, based on X-ray crystallographic analysis of the structure of reaction centers, between molecules $(BChl)_2$ and BPh. Another possibility consists in one step superexchange transfer of an electron to BPh; molecule $BChl_A$ in this case increases the connection between $^1(BChl_2)^*$ and $BChl_2^{+} \cdots BPh^{-}$. To clarify these questions it is necessary to examine exchange interactions in primary pairs. This may be done by investigating magnetic field effects and RYDMR spectra.

8.1.3. Magnetic Field Effect on Reaction Centers

The first magnetic spin effect on the isolated reaction centers of *Rodopseudomonas sphaeroides* bacteria was revealed by Blankenship et al. [4]. By using the time-resolved (in nanosecond range) optical transient absorption technique it was shown that the yield of chlorophyll triplet molecules at the excitation of samples in the magnetic field is higher compared with the case of zero field. The dependence of the transient light absorption by triplet states on the magnetic field induction looked as a step of the negative sign with a half-saturation field of about 500 G. Blankenship et al. [4] explained the results by assuming the influence of the magnetic field on the rate of singlet–triplet conversion in ion radical pairs in the reaction centers. The magnetic field effect on the fluorescence of the reaction centers that appears as a result of recombination of singlet pairs has been observed [5, 6]. An investigation of the magnetic field effect on the yield of triplets in reaction centers with removed secondary acceptors made with nanosecond time resolution has shown [7] that the yield increases and the half-saturation field decreases with time after the excitation pulse. The data obtained have permitted us to determine rate constants $k_S = 3.9 \times 10^7\ s^{-1}$ and $k_T = 7.4 \times 10^8\ s^{-1}$, and the value of exchange interaction $J = 3$ G.

Ion radicals in primary pairs in reaction centers have a fixed position like that in solids, and exchange interaction, in contrast to the cases of micelles and liquid solutions, is not averaged in time. The RYDMR method as it follows from the theory (see Section 5.3) is very sensitive to the presence of exchange interactions, and permits us to calculate that value by using the spin-locking effect.

8.1.4. RYDMR Spectra of Reaction Centers

First RYDMR spectra of reaction centers of *Rodopseudomonas sphaeroides*, R-26, with removed quinone molecules were obtained by Closs and co-workers [8] and by Norris et al. [9]. These authors have shown that the resonant microwave field increases the yield of triplet states 8% at a low value of B_1. When B_1 increases the RYDMR signal changes its polarity. This corresponds to theoretical expectations of the behavior of the RYDMR spectra in conditions when the field B_1 becomes higher than the width of the spectra determined by hyperfine interaction, by Δg, by exchange interaction, and by kinetic broadening [10, 11]. Experimental data [8, 9] lead to the next values of parameters of pairs: $0.05\ ns^{-1} \leqslant k_S \leqslant 0.09\ ns^{-1}$, $0.5\ ns^{-1} \leqslant k_T \leqslant 0.6\ ns^{-1}$, $12\ G \leqslant |J| \leqslant 20\ G$, and $40\ G \geqslant D \geqslant 6\ G$. Lersch and Michel-Beyerle [12] obtained similar values: $0.05\ ns^{-1} \leqslant k_S \leqslant 0.07\ ns^{-1}$, $0.4\ ns^{-1} \leqslant k_T \leqslant 0.6\ ns^{-1}$, and $9\ G \leqslant |J| \leqslant 13\ G$. Parameters were calculated by fitting the points from the experimental spectra by a theoretical formula (5.12) (see Section 5.3). Figure 8.2 shows the experimental RYDMR spectra obtained for reaction centers of *Chloroflexus aurautiacus* bacteria [13]. One can see the changes of the form and sign of the spectra when B_1 varies. These spectra are also described by formula (4.12) with parameters $J = 22.5 \pm 0.5$ G and $k_T = (8.0 \pm 1.5)\ 10^8\ s^{-1}$.

Another way of determining the value of the exchange interaction consists in obtaining B_1 spectra of RYDMR at constant B_0, corresponding to resonance ($B_0 = \hbar\omega/g\beta$). Figure 8.3 shows B_1 spectra obtained [13] for three different types of reaction centers. At low value of B_1 the signal, which is proportional to the yield of products produced through the triplet channel of recombination of pairs, increases with B_1. Then the signal reaches a maximum at the point where one of the triplet levels crosses the singlet level (in the rotating frame; see Section 5.2). The position of the maximum is determined by the singlet–triplet splitting at $B_1 = 0$, which is equal to exchange interaction J

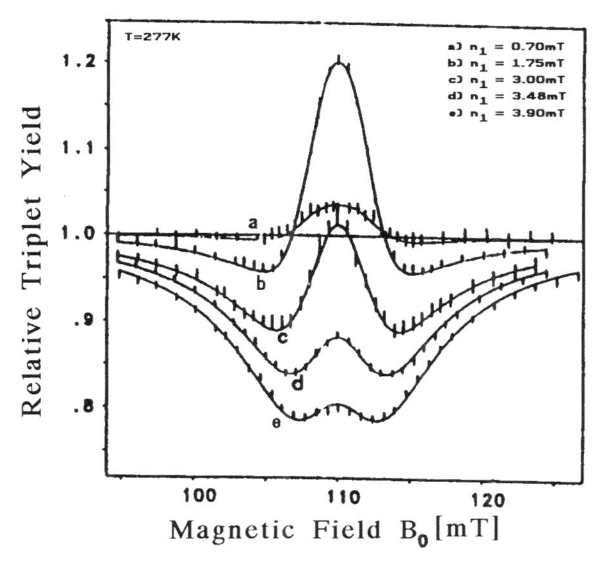

Figure 8.2 RYDMR spectra of reaction centers from *Chloroflexus aurautiacus* for different B_1 fields. (a) $B_1 = 7$ G, (b) $B_1 = 17.5$ G, (c) $B_1 = 30$ G, (d) $B_1 = 34.8$ G, (e) $B_1 = 39$ G; $T = 277$ K. The light absorbance at 865 nm was adjusted to an optical density of 1.0. The lines are theoretical fits to the data. From Lang et al. [13].

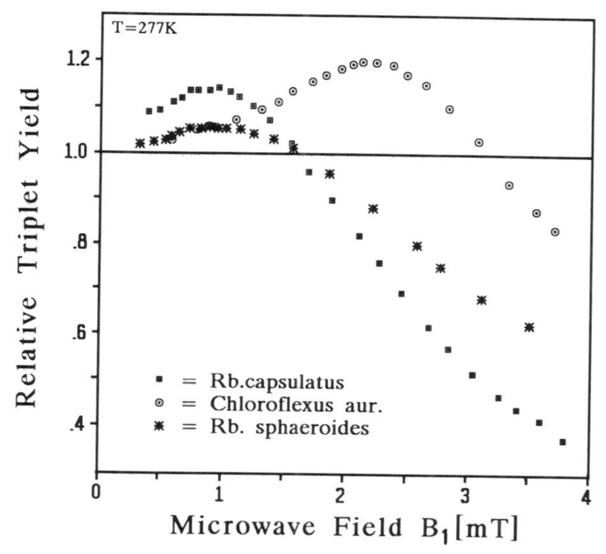

Figure 8.3 B_1 spectra of reaction centers from *Chloroflexus aurautiacus*, *Rb*, *capsulatus* wild type, and *Rb. sphaeroides*. $B_0 = 1092$ G, $T = 277$ K. From Lang et al. [13].

(if one can neglect dipole–dipole and hyperfine interactions). The position of the maximum thus gives the value of J.

Moehl et al. [14] studied the behavior of RYDMR spectra at constant B_1 varying the temperature within the range from 236 to 293 K. They were able to observe a change of the form and polarity of the signal. It is evidence of the changing value of J in relation to temperature. As was assumed [14] that this is a sequence of structural rearrangement of reaction centers. It is important to note what a correlation takes place: an increase of the exchange interaction is accompanied by a rise of the rate constant k_T. Such a correlation in variations of J and k_T corresponds, in the opinion of Lang et al. [13], to the two-step mechanism of electron transfer between $(BChl)_2$ and BPh.

Observation of the resonant change of the lifetime of the pair of ion radicals in reaction centers of *Rodopseudomonas sphaeroides* was made by Wasielewski [15]. Transient optical absorption on the wavelength at 420 nm, induced by pulse laser excitation and caused by products of reactions in pairs, was measured with a time resolution of 2.5 ns. The kinetics of the decay of induced absorption has shown the lifetime to be equal to about 20 ns. Microwaves acting at resonant conditions were observed to change the lifetime up to 25%, the sign of the effect being dependent on the value of B_1. At low B_1 (corresponding to microwave power in the pulse of 1 kW) the lifetime decreased, whereas at high B_1 (at power up to 7 kW) the lifetime was longer.

Frankevich and Pristupa [16] obtained the RYDMR spectrum by measurement of the resonant decrease of the fluorescence intensity of antenna of *Rodopseudomonas sphaeroides* bacteria. The excitation of fluorescence was produced in the absorption band of antenna carotinoids. An analysis of the shape of the spectrum has shown that the latter consists of a superposition of magnetic resonance spectra of carotinoid triplets and triplets of bacteriochlorophyll. That permitted us to calculate parameters of the fine splitting tensor for excitations involved: for the carotinoid triplet state ($D = 341 \pm 10$G, $E = 25 \pm 1$ G) and for the bacteriochlorophyll triplet state ($D = 263 \pm 10$ G, $E = 50 \pm 1$ G). The results have shown in particular that part of the singlet excitation energy from antenna is transferred to reaction centers via intermediate triplet–triplet pairs consisting of molecules of carotinoid and bacteriochlorophyll.

The examples given in this chapter demonstrate the possibilities that the RYDMR method opens for understanding electron and energy transfer mechanisms.

References

1. Govenjee, ed. *Photosynthesis*: Vol. 1: *Energy Conversion by Plants and Bacteria.* Academic Press, New York, 1982.

2. Barber, J., ed. *Primary Processes of Photosynthesis.* Elsevier, Amsterdam, 1977

3. Michel, H. *J. Mol. Biol.* **1982**, *158*, 567–562.

4. Blankenship, R. E., Schaafsma, T. J., Parson, W. W. *Biochim. Biophys. Acta* **1977**, *461*, 297–302.

5. Voznjak, V. M., Elfimov, E. I., Proskurjakov, I. I. *Dokl. Akad. Nauk SSSR* **1978**, *242*, 1200–1205.

6. Rademaker, H., Hoff, A. J., Duysens, L. N. M. *Biochim Biophys. Acta* **1979**, *546*, 248–253.

7. Ogrodnik, A., Krueger, H. W., Orthuber, H., Haberkorn, R., Michel-Beyerle, M. *Biophys. J.* **1983**, *39*, 83–87.

8. Bowman, M. K., Budil, D. E., Closs, G. L., Kostka, A. G., Wraight, C. A., Norris, J. R. *Proc. Natl. Acad. Sci. U.S.A.* **1981**, *78*, 3305–3307.

9. Norris, J. R., Bowman, M. K., Budil, D. E., Tang, J., Wraight, C. A., Closs, G. L. *Proc. Natl. Acad. Sci. U.S.A.* **1982**, *79*, 5532–5536.

10. Tang, J., Norris, J. R. *Chem. Phys. Lett.* **1983**, *94*, 77–80.

11. Lersch, W., Michel-Beyerle, M. E. *Chem. Phys. Lett.* **1987**, *136*, 346–351.

12. Lersch, W., Michel-Beyerle, M. E. RYDMR—theory and application. In *Advanced EPR in Biology and Biochemistry*; Hoff, A. J., ed. Elsevier, Amsterdam, 1989, p. 685.

13. Lang, E., Lersch, W., Tapperman, P., Coleman, W. J., Youvan, D. C., Feick, R., Michel-Beyerle, M. E. *Current Res. Photosyn.* **1990**, *1*, 137.

14. Moehl, K. W., Lous, E. J., Hoff, A. J. *Chem. Phys. Lett.* **1985**, *121*, 22–27.

15. Wasielewski, M. R., Boek, C. H., Bowman, M. K., Norris, J. R. *J. Am. Chem. Soc.* **1983**, *105*, 2903–2904.

16. Frankevich, E. L., Pristupa, A. I. *Izvestia Akad. Nauk SSSR, Ser. Fiz.* **1988**, No. 2, 222–225.

CHAPTER

9

Radio-Induced Magnetic Isotope Effect

The microwave pumping of the ESR transitions modifies the rate of spin evolution of the radical pairs (RPs) and, consequently, their chemistry, resulting in RYDMR, a new principle of frequency-tuned resonant chemical reception of microwaves (mw). However, this reception is nonselective because all the RPs are equally subjected to microwave perturbation. In this chapter as well as in the next one the mw pumping which is selective with respect to nuclear spins, magnetic moments, and their orientations will be discussed.

The magnetic isotope effect (MIE) [i.e., the dependence of chemical reactivity of radicals and RPs on their nuclear magnetic properties (see Chapter 3)] results in the separation of the magnetic and nonmagnetic isotopes. A new way to increase the MIE and isotope separation efficiency is the selective mw pumping of RPs with magnetic nuclei, which accelerates the spin conversion of these pairs and stimulates their intrapair reactions.

The idea of the radiowave-induced magnetic isotope effect (RIMIE) was clearly formulated in 1981 [1] and elegantly embodied in 1991 [2] in the photolysis of dibenzylketone (DBK) in the sodium dodecyl sulfate (SDS) micelles. The reaction scheme of photolysis was discussed in Chapter 3.

The laser pulses used for DBK photolysis (388 nm wavelength, 15 ns duration, 50 mJ energy) were synchronized with mw pulses. Their duration (2 μs) was chosen to exceed by at least an order of magnitude the lifetime of the RP in the micelle, 150 ns. The time interval between each synchronized couple of laser and mw pulses was 100 ms (the sequence of pulses is shown schematically in Fig. 9.1).

The frequency of the mw irradiation was 1530 MHz (magnetic field strength

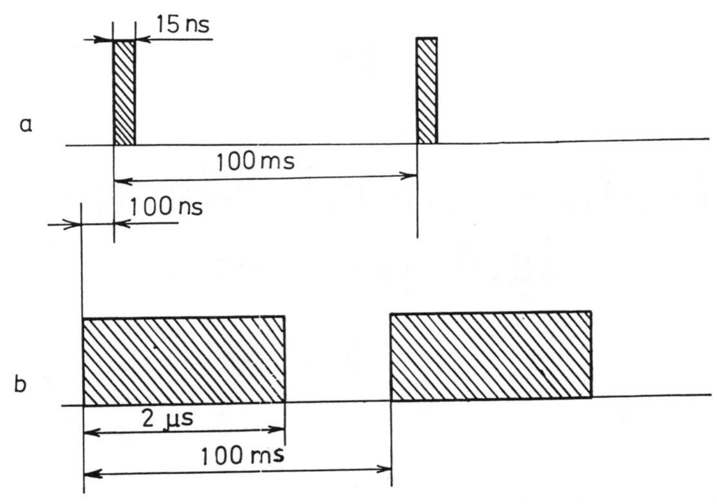

Figure 9.1 The scheme of pulse sequences and durations: laser (a) mw (b) pulses.

603 G). It corresponds to the pumping of the high field ESR transition in $PhCH_2{}^{13}CO$ radical and is equivalent to the stimulation of the triplet–singlet conversion of the magnetic RPs ($Ph\dot{C}H_2\ {}^{13}\dot{C}OCH_2Ph$), while the nonmagnetic pairs ($Ph\dot{C}H_2\ {}^{12}\dot{C}OCH_2Ph$) remain unperturbed.

The DBK isotope enrichment and its chemical conversion are related by the equation

$$\ln S = (1 - \alpha) \ln(1 - f^*) \tag{9.1}$$

where $S = \delta/\delta_0$, δ_0 and δ are, respectively, the ^{13}C content in the starting ketone and that after photolysis, f^* is the ^{13}C-DBK chemical conversion. The coefficient α shows the efficiency of the isotope separation (see Chapter 3).

The experimentally measured DBK isotope enrichment S as a function of $(1 - f^*)$ is depicted in Fig. 9.2 and conforms to Eq. (9.1): the higher the ketone conversion the higher the magnetic isotope enrichment of its remains. The remarkable effect is that the isotope separation coefficients are different: for mw pumped photolysis $\alpha_{mw} = 1.25 \pm 0.01$, while for photolysis without pumping $\alpha_0 = 1.21 \pm 0.01$. In fact, the mw pumping turns out to increase the RP isotope selectivity and the ketone isotope enrichment.

The coefficient α in Eq. (9.1), as shown in Chapter 3, can be expressed in terms of the recombination probabilities of the magnetic and nonmagnetic pairs, P^* and P, respectively:

$$\alpha = (1 - P)/(1 - P^*)$$

Let P_0^* and P_{mw}^* denote the recombination probability of unperturbed magnetic RP and that of magnetic RP subjected to mw pumping. Then

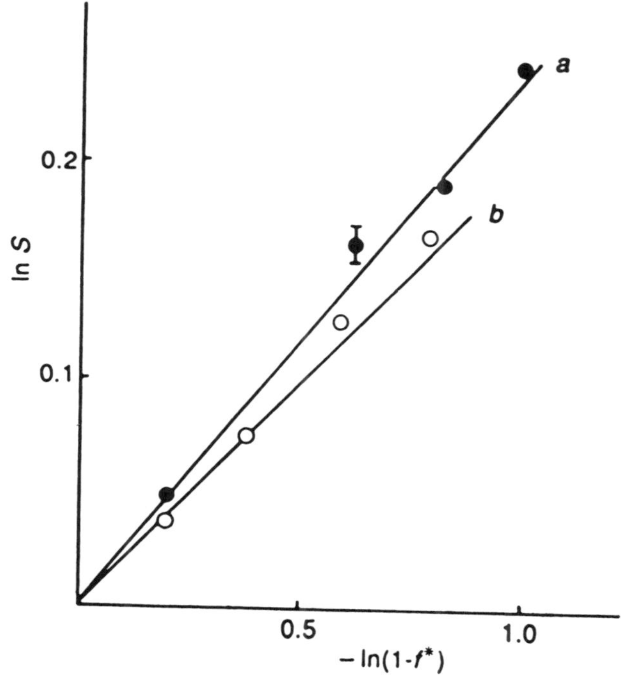

Figure 9.2 Isotope enrichment S of dibenzylketone as a function of ^{13}C-DBK chemical conversion f^*: (a) mw pumping, $B_1 = 0.3$ mT; (b) no pumping [2].

$$\alpha_0 = \frac{1 - P}{1 - P_0^*} \tag{9.2}$$

$$\alpha_{mw} = \frac{1 - P}{1 - (\frac{1}{2}P_0^* + \frac{1}{2}P_{mw}^*)} \tag{9.3}$$

Equation (9.3) corresponds to the experimental conditions when the subensemble of the RPs with the α_n spins of ^{13}C nuclei evolves from the T to the S state naturally, under the action of the native HFI, while another one with β_n spins evolves under the action of both native HFI and mw pumping. (Let us remind the reader that only high field ESR line, i.e., the transitions within the subensemble of magnetic RPs with β_n nuclear spin of ^{13}C in $PhCH_2{}^{13}CO$ radical, is pumped.) Then, instead of P_0^* in Eq. (9.2) the sum $(\frac{1}{2}P_0^* + \frac{1}{2}P_{mw}^*)$ should be taken in Eq. (9.3). It is evident that P_{mw}^* is the total recombination probability of the magnetic pair if both its nuclear spin subensembles are subjected to mw pumping.

By combining (9.2) and (9.3) one can derive

$$\frac{2\alpha_0}{\alpha_{mw}} = 1 + \frac{1 - P_{mw}^*}{1 - P_0^*} \tag{9.4}$$

or

$$\frac{2\alpha_0}{\alpha_{mw}} - 1 = \frac{1 - P^*_{mw}}{1 - P^*_0} = 0.936 \tag{9.5}$$

To obtain P^*_{mw} it is necessary to know P^*_0. The latter is not known accurately, however, the value of 0.50 can be used with confidence since it is derived from the comparison of recombination probabilities and their magnetic field dependencies for the RPs chemically and magnetically similar to those arising in DBK photolysis.

Taking this value for P^*_0 and assuming that P^*_{mw} consists of the two contributions

$$P^*_{mw} = P^*_0 + P_{mw}$$

where the first is due to the spin evolution induced by HFI and the second determines the net contribution of the mw pumping, one can find the ratio

$$P_{mw}/P^*_0 \approx 6 \times 10^{-2}$$

Thus, the net share of mw pumping into the recombination probability of the magnetic pair is about 6% (at B_1, the mw irradiation amplitude, of 0.3 mT).

It is worth estimating the time scale for different events in the RPs spin and chemical evolution. The RP lifetime is determined by the rate constant of the acyl radical decarbonylation and is equal to 150 ns. The time of T_0–S spin conversion is about 2 ns and much shorter than that between the radical reencounters in pair, $\tau_{enc} = 36$ ns. The intrinsic time of the mw-induced electron spin inversion, $\tau_{mw} = (\gamma B_1)^{-1} = 120$ ns, corresponds to that of the equilibrating $T_\pm$ and T_0 state spin populations due to mw pumping. This time is considerably longer than the time interval between radical reencounters and much longer than the T_0–S conversion time.

For the mw pumping to be efficient in the RP triplet–singlet conversion and isotope separation, the mw-induced $T_\pm$–T_0 transitions should occur within the time interval τ_{enc} between the radical reencounters. In other words, the requirement $\tau_{mw} \leqslant \tau_{enc}$ should be satisfied. The reverse relation between these times might mean that the mw pumping would not have enough time to complete the $T_\pm$–T_0 transitions and effectively interfere in the triplet–singlet RP evolution. In the experiments described above this unfavorable relation, $\tau_{mw} > \tau_{enc}$, holds and this is likely the main reason for the relatively low efficiency of mw pumping in isotope selection.

In view of these arguments, the spin locking regime for the nonmagnetic pairs ($Ph\dot{C}H_2$ $^{12}\dot{C}OCH_2Ph$) is anticipated to be highly promising for isotope selection, since it prevents their recombination and, consequently, increases the coefficient of the isotope selection. The most efficient would be the combination of both regimes—spin locking for the nonmagnetic pairs and microwave pumping for the magnetic ones.

The RIMIE effect is the first demonstration of the net isotope selectivity induced by mw pumping of radical pairs. It was detected even earlier, in 1988, for the H/D isotope pair [4], but only in partly selective mw pumping. Okazaki et al. [4] studied the photoreduction of menadione (MD), 2-methyl naphthoquinone, in SDS micelles composed of the SDS molecules (SDS-H) perdeuterated SDS molecules (SDS-D), and their mixtures. The photoinduced reaction mechanism includes the hydrogen (or deuterium) atom abstraction from SDS-H (or SDS-D) molecules by the excited triplet state of MD [5].

The scheme of the RP reactions is given below:

$$ {}^{3}\mathrm{MD} + \mathrm{SDS} \rightarrow (\mathrm{M\dot{D}H\ S\dot{D}S})^{\mathrm{T}} \begin{cases} \rightarrow \text{coupling products} \\ \rightarrow \mathrm{M\dot{D}H} + \mathrm{S\dot{D}S} \end{cases} $$

The triplet RP $(\mathrm{M\dot{D}H\ S\dot{D}S})^{\mathrm{T}}$ composed of $\mathrm{S\dot{D}S}$ ($\mathrm{S\dot{D}S}$-H or $\mathrm{S\dot{D}S}$-D) radical and semiquinone $\mathrm{M\dot{D}H}$ radical either reacts after triplet–singlet conversion to form coupling products or dissociates to free $\mathrm{M\dot{D}H}$ and $\mathrm{S\dot{D}S}$ radicals. Only the $\mathrm{S\dot{D}S}$ is scavenged by water-soluble nitroso spin trap perdeuteriodimethylnitrosobenzene sulfonate (TNO) because the reactivity of the spin trap to the semiquinone radical is very low. The stable spin adduct, nitroxide radical $\mathrm{T(SDS)N\dot{O}}$, was measured by ESR and its chemical yield was detected as a measure of the competition between the radical coupling reaction in pair and the radical escape.

The yield of spin adduct is sensitive to mw pumping and is shown to decrease under its influence in accordance with the reaction scheme. The mw pumping accelerates the triplet–singlet conversion of the RP and stimulates the coupling cage reactions, suppressing the competing radical escape processes.

The mw modulation of the spin adduct yield as a function of magnetic field is shown in Fig. 9.3. These dependencies are, as a matter of fact, the ESR spectra of the RPs ($\mathrm{M\dot{D}D\ S\dot{D}S}$-D) and ($\mathrm{M\dot{D}H\ S\dot{D}S}$-H), corresponding to the superposition of the ESR spectra of RP partners. More exactly, they are the RYDMR spectra detected by trapping of the free radicals that avoid the reaction in RP.

One may predict that one of the isotopes, H or D, may be enriched by mw pumping in the mixed micelles of SDS-H and SDS-D molecules taken in a 1 : 1 ratio, if the magnetic field is chosen properly. For example, at the magnetic field 338.5 mT, indicated in Fig. 9.3 by a vertical arrow, the mw irradiation was found to cause a decrease of the spin adduct yield by about 15% for SDS-D radicals and by only about 5% for SDS-H radicals. It means that under mw pumping the SDS-H radicals are able to avoid the coupling reactions in RPs and be transferred into the spin adduct product in comparison with SDS-D radicals, which are subjected to mw pumping to a greater extent and are forced to react in the pair. It results in hydrogen isotope enrichment of the spin adduct with respect to deuterium.

In contrast to the $^{13}\mathrm{C}/^{12}\mathrm{C}$ RIMIE, in this case the isotope selectivity of microwave pumping is restricted due to the superposition of the ESR transitions in both isotope pairs ($\mathrm{M\dot{D}D\ S\dot{D}S}$-D) and ($\mathrm{M\dot{D}H\ S\dot{D}S}$-H). The best selectivity is achieved if their ESR transitions are separated on a frequency

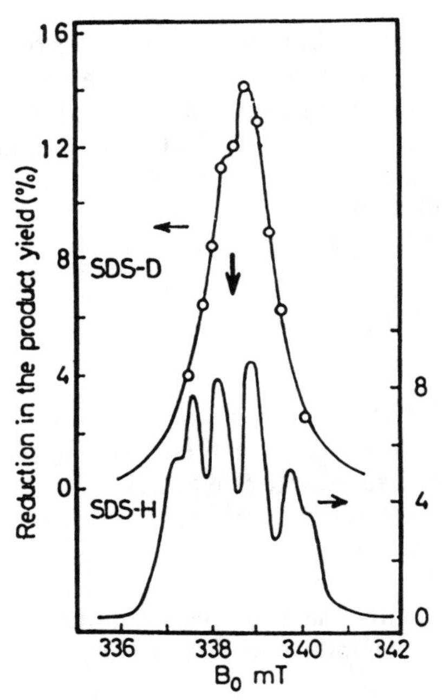

Figure 9.3 The reduction in product yield (RPY) induced by mw irradiation in the photoreduction of menadione in SDS-H and SDS-D micelles as a function of magnetic field B_0. The RPY dependencies on B_0 are in fact the ESR spectra of RPs in micelles. The vertical arrow indicates the magnetic field chosen for mw pumping.

scale. Then it would be available to pump isotopically different radical pairs independently. This was the case for dibenzylketone photolysis discussed previously.

In conclusion, it should be noted that RIMIE as a new type of isotope phenomena is a beautiful part of the magnetic microwave scenario of chemical reaction and, moreover, it can be used to increase the MIE-induced isotope separation.

References

1. Buchachenko, A. L., Tarasov, V. F. *Russ. J. Phys. Chem.* **1981**, *55*, 936.

2. Tarasov, V. F., Bagryanskaya, E. G., Grishin, Yu. A., Sagdeev, R. Z., Buchachenko, A. L. *J. Chem. Soc. Mendeleev Commun.* **1991**, *1*, 85.

3. Buchachenko, A. L. Magnetic isotope effect and isotope selection in chemical reactions. In *Progress in Reaction Kinetics*, Jennings, K. R., Cundall, R. B., Margerum, D. W., eds. Pergamon Press, Oxford, 1984, Vol. 13, no. 3, p. 164.

4. Okazaki, M., Shida, T., Sakata, S., Konaka, R., Toriyama, K. *J. Phys. Chem.* **1988**, *92*, 1402.

5. Okazaki, M., Sakata, S., Konaka, R., Shida, T. *J. Chem. Phys.* **1987**, *86*, 6792.

CHAPTER

10

Microwave-Stimulated Nuclear Polarization

Stimulated nuclear polarization (SNP) is a result of microwave pumping the ensembles of radical pairs with selected orientations of the nuclear spins. This phenomenon is similar to CIDNP (see Chapter 3); the only difference is that in contrast to CIDNP where the nuclear spin selection of pairs via triplet–singlet conversion is accomplished by native, existing in radicals, magnetic interactions, in SNP the selection is produced by the resonant microwave magnetic field. In other words, the RP again serves as the selective chemical microwave receiver, but now the indicator of the reception is the nuclear polarization in the reaction products.

In chemical reactions subjected to microwave (mw) resonant and, hence, nuclear spin selective pumping both phenomena, native CIDNP and SNP, coexist, so that the experimentally measured NMR spectrum of the product molecules inherits both contributions. The best way to discriminate between them is to perform the pumping and detection separately, both in time and space.

A block diagram of the experimental setup, home-made and used by Sagdeev and Bagryanskaya [1], the pioneers of SNP studies, is shown in Fig. 10.1.

The chemical reagent solution is irradiated by laser or by another light source in the cavity placed in the magnetic field B_0 of the auxiliary magnet and the mw field B_1 is applied perpendicular to the B_0 direction. The flow system transfers the chemically reacting solution into the NMR probe to detect the SNP spectra of reaction products. To synchronize the mw generator and laser with the NMR spectrometer the system is equipped with a controlling

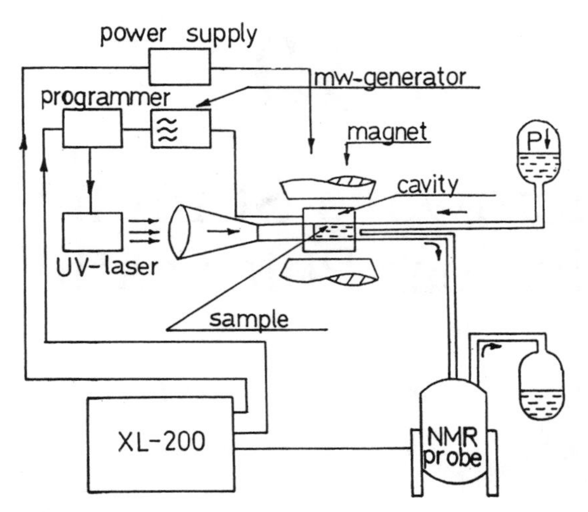

Figure 10.1 A schematic diagram of the setup for SNP detection [1].

programmer. The net SNP spectrum is recorded as the difference between the NMR signals with and without mw pumping. This procedure abolishes the CIDNP contribution and extracts the net nuclear polarization induced by mw pumping.

The setup allows us to accomplish two types of experiments. First, the NMR signal intensities monitoring as a function of B_0 and detection of SNP spectra, and, second, the variation of the time delay τ_d between laser and mw pulses at fixed B_0. The latter, the time-resolved regime, makes it possible to trace the dynamic behavior of radicals, radical pairs, and other short lived paramagnetic intermediates. The τ_d values cover the time domain from 10 ns to 75 μs at the maximal mw pulse duration of 30 μs. Although the time resolution in the SNP experiments is limited by the steepness of the head and tail profiles of mw pulse, the Sagdeev and Bagryanskaya [1] have been able to reduce the dead transient time to 30 ns.

10.1. SNP in High Magnetic Fields

Figure 10.2 clearly illustrates the background physics of the SNP phenomenon. Suppose there is a triplet benzoyl-benzyl RP generated by the photolysis of dibenzylketone with ^{13}C in the carbonyl group. ESR spectrum of the pair consists of a central group of lines that belongs to the $PhCH_2$ radical (not shown in Fig. 10.2) and of the $PhCH_2{}^{13}CO$ radical doublet with hyperfine splitting $a(^{13}C)$. Since $a(^{13}C)$ is known to be positive, the low-field line of the doublet is attributed to the ensemble of radical pairs with ^{13}C nuclei oriented up (α_n spins populating the lower nuclear Zeeman level), while the high-field

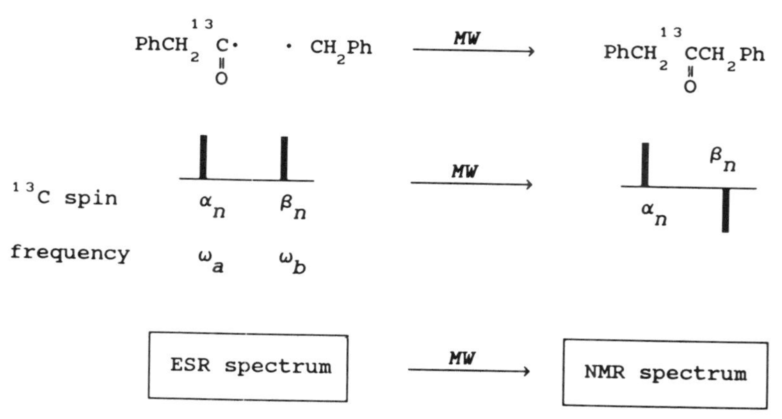

Figure 10.2 Schematic illustration of SNP origin. The selective mw pumping the ^{13}C RP ensembles with α_n and β_n nuclear spins at frequencies ω_a and ω_b accelerates their triplet–singlet conversion and transforms the ESR spectrum of the pair into the NMR spectrum of the product molecule. The latter inherits the nuclear spin orientations of its precursor, the mw pumped RP.

line belongs to that with ^{13}C nuclei oriented down (β_n spins occupying the upper nuclear Zeeman level).

The fast intersystem T_0–S conversion induced by HFI equalizes the populations of these two spin states. The mw pumping of the ESR transition in the first ensemble of pairs at frequency ω_a stimulates $T_{\pm}$–T_0 spin conversion and additionally populates the T_0 state of the pairs with nuclear spins α_n resulting in the acceleration of their recombination and dominant generation of the molecules with α_n spins (the positive SNP). The mw pumping of the second RP ensemble with β_n nuclear spins at frequency ω_b selectively produces dibenzylketone molecules with ^{13}C nuclear β_n spins (the negative SNP). As a result, the mw pumping transforms the ESR spectrum of the radical pair into the NMR spectrum of the molecule with polarized nuclei.

These predictions are perfectly confirmed experimentally (Fig. 10.3).

The combination of the absorption–emission (A/E) SNP spectrum lines is an indication of sorting the nuclear orientations and evidence of the leading role of T_0–S conversion in RP spin evolution. It is easy to predict that the SNP spectrum should be inverted, E/A instead of A/E, if either the sign of the hyperfine coupling constant is altered (and, therefore, nuclear spin orientations become inverted), or the spin multiplicity of the starting spin state of the pair is changed resulting in the inversion of the spin evolution direction.

Both predictions were experimentally confirmed. For instance, the SNP spectrum detected by NMR of duroquine (DQ, tetramethyl-*p*-benzoquinone) CH_3 protons at the photolysis of DQ in methanol shows an A/E pattern (Fig. 10.4a). This feature is unambiguously assigned to the radical pair $(\dot{D}QH\ \dot{D}QH)^F$, which results from the encounters of freely diffusing radicals DQH, the intermediates of DQ photoreduction. The spin behavior of such a

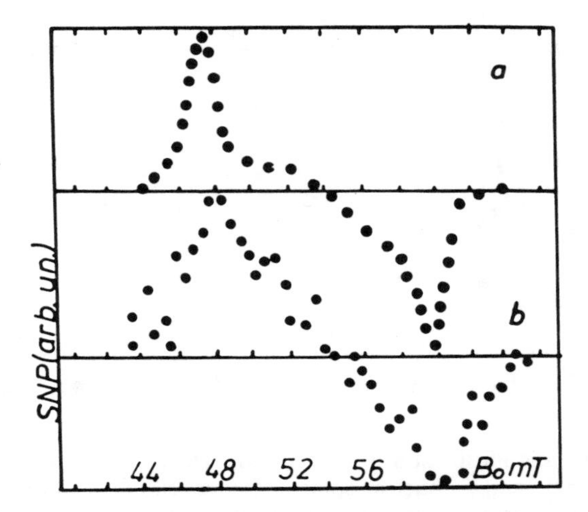

Figure 10.3 SNP spectra detected for carbonyl ^{13}C of dibenzylketone molecules at the photolysis in SDS micelles (a) and in benzene (b); ESR transition frequency is 1530 MHz, $B_1 = 0.2$ (a) and 0.6 mT (b). The former spectrum is recorded for ketone with natural ^{13}C abundance, and the latter for the 45% enriched ketone [2].

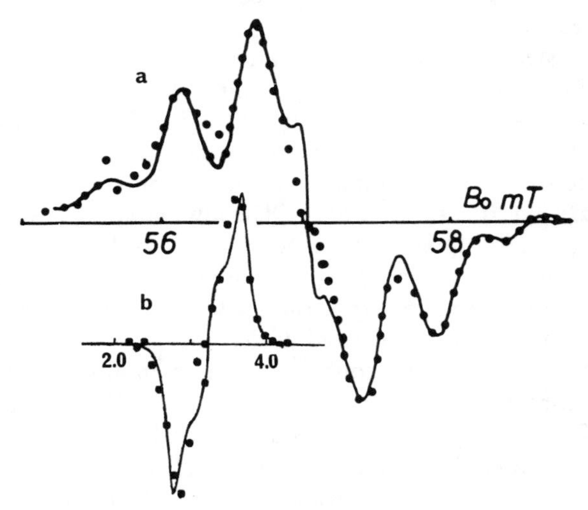

Figure 10.4 SNP spectra detected for CH_3 protons of duroquinone (tetramethyl-*p*-benzoquinone) (a) and for ring protons of benzoquinone (b) [3].

pair is known to be similar to that of the triplet one. However, the SNP spectrum detected by NMR of benzoquinone (Q) protons exhibits an E/A pattern (Fig. 10.4b) reflecting the nuclear spin population induced in $(\dot{Q}H\ \dot{Q}H)^F$ RP by mw pumping. The only difference is the opposite sign of HFI constants in radicals DQH and QH, positive for the CH_3 protons of DQH and negative for the ring protons of QH. The inversion of the HFI sign is immediately followed by phase inversion in the SNP spectra.

SNP linewidth as a function of B_1 was studied by Meng et al. [4] in the photolysis of Q in the presence of isopropanol and the lifetime of the RP was estimated from this dependence as 20–30 ns.

The photoisomerization of fumaronitrile FN (*trans*-isomer) into maleonitrile MN (*cis*-isomer) sensitized by naphthalene in acetonitrile results in an SNP spectra with opposite phases, A/E and E/A, for FN and MN (Fig. 10.5). The phase inversion evidences the different spin multiplicities of the ion radical pairs, $(\dot{F}\bar{N}\ \dot{Naph}^{+})^S$ and $(\dot{M}\bar{N}\ \dot{Naph}^{+})^T$, which participate as photolysis intermediates, in agreement with the conclusion independently derived from the CIDNP study of this reaction [5].

The photoreaction of anthraquinone (A) with triethylamine (TEA) in methanol, acetonitrile, and benzene provides another example of the identification of intermediate short-lived pairs [6]. SNP spectra for CH_2 protons of TEA in methanol demonstrate the A/E phase, while for the methanol hydroxyl proton the SNP phase is inverted, E/A (Fig. 10.6). It unambiguously proves the ion radical mechanism of the reaction:

$$A + TEA \rightleftharpoons (\dot{A}^- \ TE\dot{A}^+)$$

$$\dot{A}^- + CD_3OD \rightarrow \dot{A}D + CD_3O^-$$

$$\dot{N}^+(CH_2CH_3)_3 + CD_3O^- \rightarrow CH_3\dot{C}HN(CH_2CH_3)_2 + CD_3OH$$

The primary step is the reversible photo-induced electron transfer accompanied

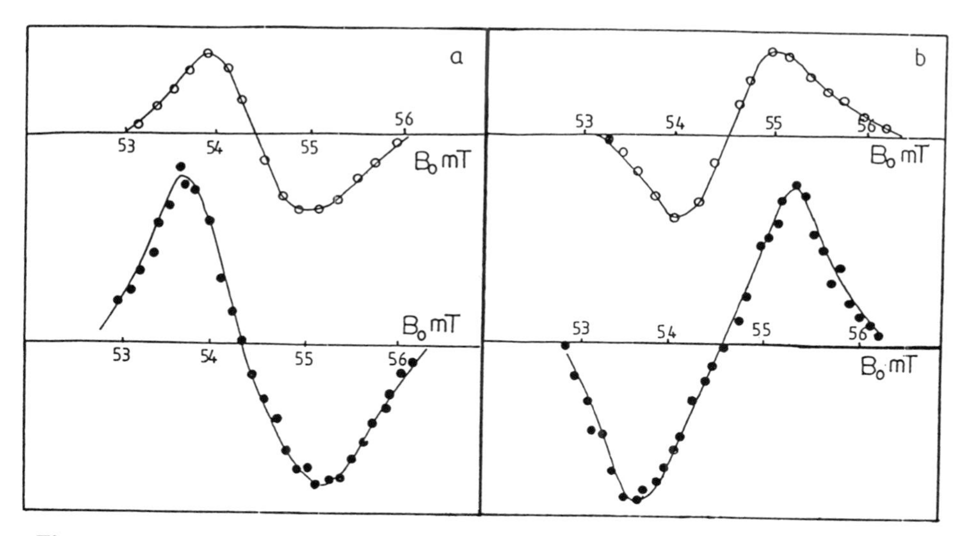

Figure 10.5 SNP spectra detected for protons of fumaronitrile (a) and maleonitrile (b) in the photolysis of fumaronitrile sensitized by naphthalene [1]; $B_1 = 0.1$ mT (top) and 0.25 mT (bottom).

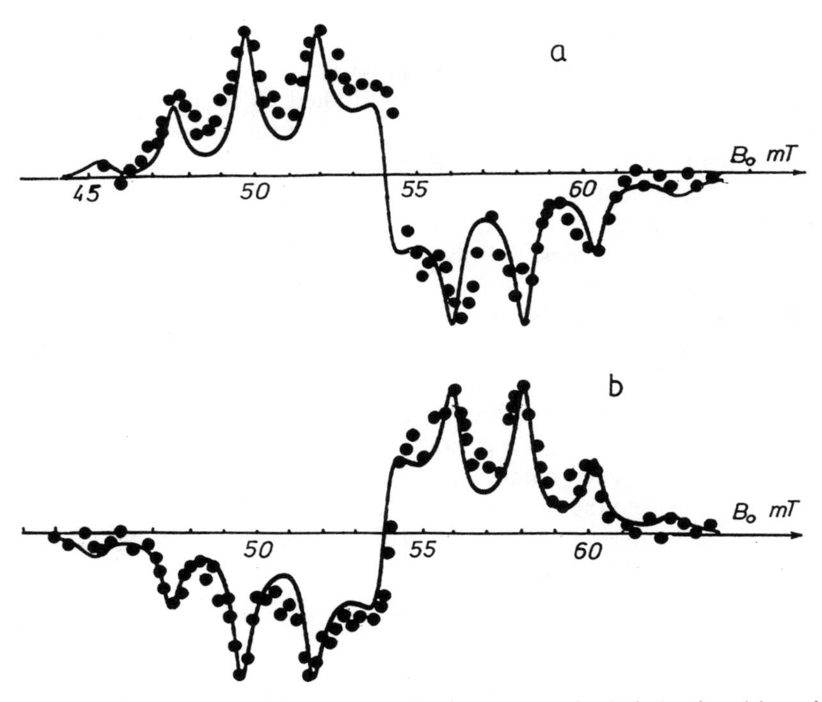

Figure 10.6 SNP spectra detected for CH_2 protons of triethylamine (a) and for methanol hydroxyl protons (b) in the photoreaction of anthraquinone with triethylamine in acetonitrile and methanol (a) and in methanol (b) [6].

by the generation of TEA molecule with polarized CH_2 protons (Fig. 10.6). The SNP spectra for CH_3 protons are not expected to appear, since in the $\dot{TEA}^+$ radical the HFI for these protons is negligibly small and noneffective either in CIDNP or in SNP. Hydroxyl methanol proton originates from the CH_2 group of the $\dot{TEA}^+$ radical, which avoids the reaction in the pair and, hence, carries the opposite SNP phase inherited from this radical.

In benzene the SNP spectrum is different from that in methanol and acetonitrile in accordance with the change of the reaction mechanism: the hydrogen atom transfer from TEA to excited triplet A molecules in benzene replaces the electron transfer in polar solvents. The SNP spectrum (Fig. 10.7) inherits the nuclear spin population created by mw pumping of the triplet RP [$\dot{A}H$ $CH_3\dot{C}HN(CH_2CH_3)_2$] with large HFI for CH_2 protons typical for such sorts of radicals.

The essential requirement for SNP spectra detection is that mw field amplitude $\omega_1 = \gamma_e B_1$ should be low enough to satisfy the inequality $\omega_1 < |\omega_a - \omega_b|$, which implies the separate pumping of individual ESR transitions in RP partners. In the opposite case of $\omega_1 > |\omega_a - \omega_b|$ the mw field affects both radicals in pair. It synchronizes their electron spin precessions and slows down the rate of triplet–singlet conversion resulting in an SNP spectrum that

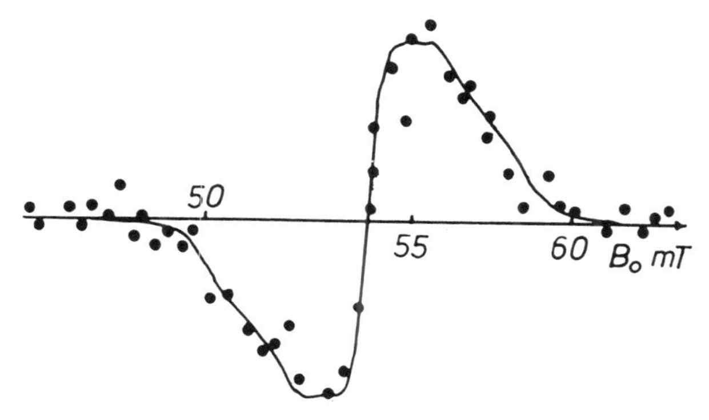

Figure 10.7 SNP spectrum detected for CH_3 protons of triethylamine in the photoreaction of anthraquinone with triethylamine in benzene [6].

is phase inverted relative to that detected at $\omega_1 < |\omega_a - \omega_b|$. It is justified by Fig. 10.8, which depicts two phase-inverted SNP spectra recorded at different $\omega_1 : \omega_1 < |\omega_a - \omega_b|$and $\omega_1 > |\omega_a - \omega_b|$. The latter relation corresponds to the spin-locking effect discussed in Chapter 5.

Salikhov and co-workers [8] have formulated the following rule to predict the SNP sign

$$\Gamma = \mu \varepsilon a \psi \eta$$

similar to that for CIDNP suggested by Kaptein [9]. Here $\mu = 1$ or -1 for

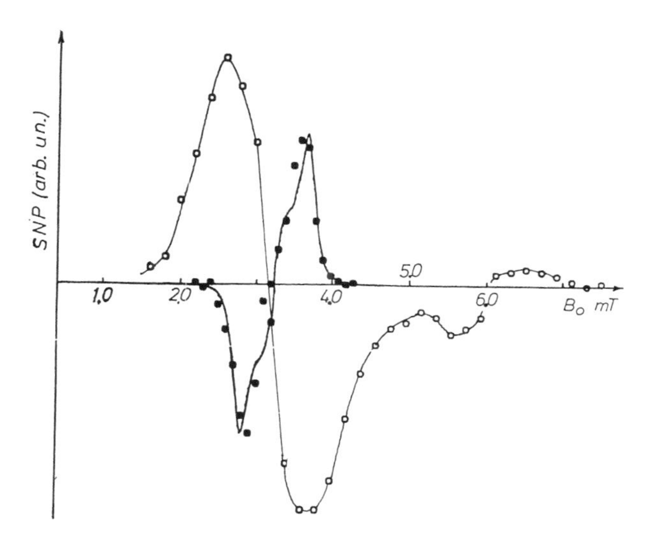

Figure 10.8 SNP spectra detected for the ring protons of benzoquinone during the photolysis of benzoquinone in acetonitrile; $B_1 = 0.07$ mT (filled circles), $B_1 = 1$ mT (open circles). The phase inversion of the latter spectrum corresponds to spin-locking effect at $B_1 = 1$ mT [7].

triplet or singlet RP, $\varepsilon = 1$ or $= -1$ for cage or out-of-cage products, respectively, $a = 1$ or $= -1$ for positive or negative HFI constants, ψ is 1 or -1 depending on whether mw pumping is applied to the low- or to high-field component of the ESR spectrum, $\eta = 1$ for the normal SNP regime but $\eta = -1$ for the spin-locking effect, and $\Gamma > 0$ or $\Gamma < 0$ corresponds to absorption or emission in SNP spectra.

10.2. SNP in Biradicals

The biradical as a molecular mw receiver is similar to RP only if the exchange interaction J between the radical centers is vanishingly small. If it is not the case the behavior of the biradicals and their SNP spectra is dictated by the relation between J and Zeeman energy $g\beta H$.

Figure 10.9 illustrates the level diagram of the biradical at the typical ratios of J and $g\beta H$ values. At $|2J| \ll g\beta H$ hyperfine coupling mixes only S and T_0 levels, so that the spin-sorting process in RP prevails and the SNP spectra exhibit A/E (or E/A) patterns according to the scheme presented in Fig. 10.2. When $|2J| \approx g\beta H$ the dominant contribution in both CIDNP and SNP is supplied by the HFI-induced T_-–S transitions. The mw pumping populates the T_- spin substates (Fig. 10.9) increasing the T_-–S flux of molecules with negatively polarized nuclei, since only $T_-\alpha_n \rightarrow S\beta_n$ transitions are allowed (Chapters 2 and 3). Finally, at $|2J| \gg g\beta H$ the T and S terms are far distant and both CIDNP and SNP are absent, because neither HFI nor $g\beta H$ is powerful enough to induce transitions between strongly energy-separated levels. Of course, the schemes shown in Fig. 10.9 predict only the trends in the character of SNP spectra as a function of J, $g\beta H$, and HFI rather than their quantitative behavior, which should be calculated rigorously.

Excellent illustrations of the nuclear spin selective mw pumping of biradicals were given by Koptyug et al. [10]. They investigated the SNP spectra in the cyclic ketone photolysis, which generates the triplet biradicals with polymethy-

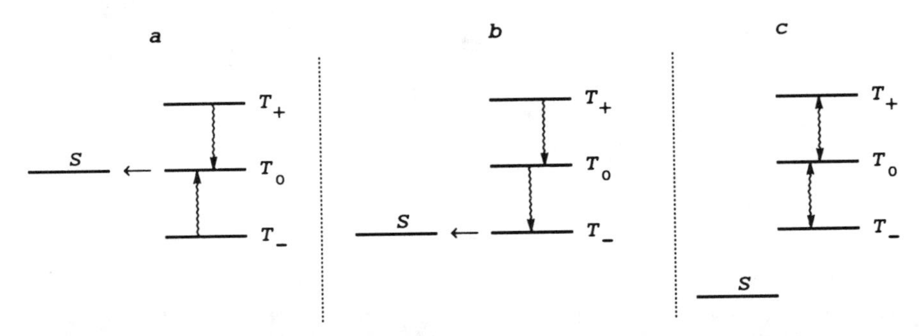

Figure 10.9 The structure of spin state levels in biradicals with different relations between exchange and Zeeman energy: $|2J| \ll g\beta H$ (a), $|2J| \approx g\beta H$ (b), $|2J| \gg g\beta H$ (c). The wavy lines indicate the mw-induced transitions.

lene chains of different size and, therefore, of varying exchange interaction J between the alkyl and acyl radical centers:

$$O{=}\dot{C}{-}(CH_2)(CH_2)_7\dot{C}H_2 \quad (1)$$

$$O{=}\dot{C}{-}(CH_2)(CH_2)_8\dot{C}H_2 \quad (2)$$

$$O{=}\dot{C}{-}(CH_2)(CH_2)_9\dot{C}H_2 \quad (3)$$

$$O{=}\dot{C}{-}(CH_2)(CH_2)_{10}\dot{C}H_2 \quad (4)$$

$$O{=}\dot{C}{-}(CD_2)(CH_2)_9\dot{C}D_2 \quad (5)$$

$$O{=}\dot{C}{-}C(CH_3)_2(CH_2)_9\dot{C}(CH_3)_2 \quad (6)$$

The SNP spectra of ketones and saturated aldehydes arising from intramolecular coupling reactions of the biradicals *2–6* in low magnetic field with 308 MHz pumping frequency are shown in Fig. 10.10. They are very similar and evidence the dominant contribution of the $T_{-}\alpha_n \rightarrow S\beta_n$ transitions to SNP in agreement with the condition $|2J| \approx g\beta H$ (Fig. 10.9b). The only difference is that the maximum of stimulated polarization that exactly satisfies the equality $|2J| = g\beta H_{max}$ is shifted with respect to magnetic field for various biradicals and this shift is a measure of the conformationally averaged values of exchange energy as a function of the size of molecular bridge between acyl–alkyl radical centers.

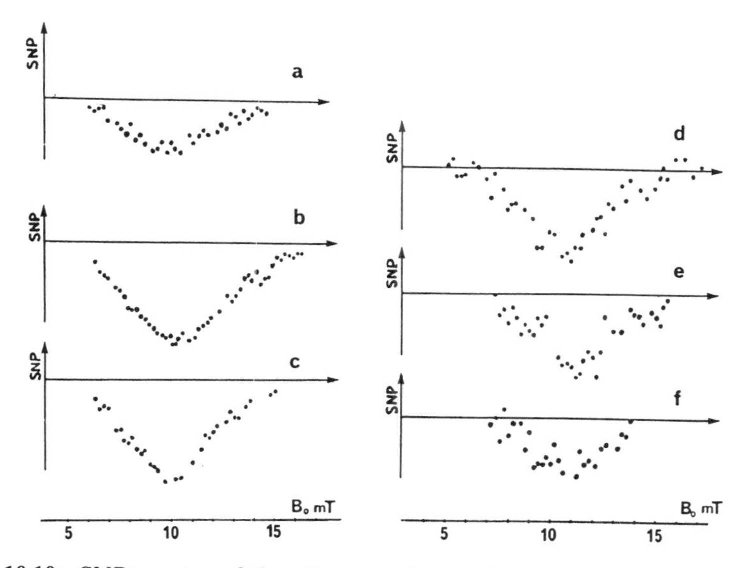

Figure 10.10 SNP spectra of the photoreaction products generated from biradicals: ketones from *3*, detected by NMR of γ-protons (a); from *2*, detected by NMR of γ-protons (b); from *4*, detected by NMR of γ-protons (c); from *5*, detected by ^{13}C NMR of α,α' carbon atoms (d); unsaturated aldehyde from *6*, detected by NMR line $\delta = 1.7$ ppm (e); from *6*, detected by NMR line $\delta = 4.5$ ppm (f). The frequency of pumping $f = 308$ MHz, $B_1 = 0.75$ mT (a–c), 1.0 mT (d), 0.9 mT (e, f) [10].

SNP spectra in high magnetic fields (the frequency of pumping 1531 MHz) are shown in Fig. 10.11. The SNP spectrum of ketone boron from biradical *2* corresponds to the predominance of $T_{-}\alpha_n \rightarrow S\beta_n$ transitions (i.e., even in high field the ratio $|2J| > g\beta H$ remains valid for this biradical). However, the SNP spectrum of ketone from biradical *3* reveals an A/E doublet instead of the T_0–S transition as the main contributor to SNP, while in the weak field the SNP spectrum is attributed to the $T_{-}\alpha_n \rightarrow S\beta_n$ transitions in the same biradical (Fig. 10.10a). The reason is that the magnetic field removes the T_{-} level far from T_0 one, and, hence, the contribution of the direct $T_{-}\alpha_n \rightarrow S\beta_n$ transitions can be ignored so that the coexisting T_0–S conversion predominates.

Thus, SNP is a physical test to analyze interradical exchange interaction and SNP spectra of molecules born from biradicals appear to be delicate probes for the sign and magnitude of this radical–radical coupling.

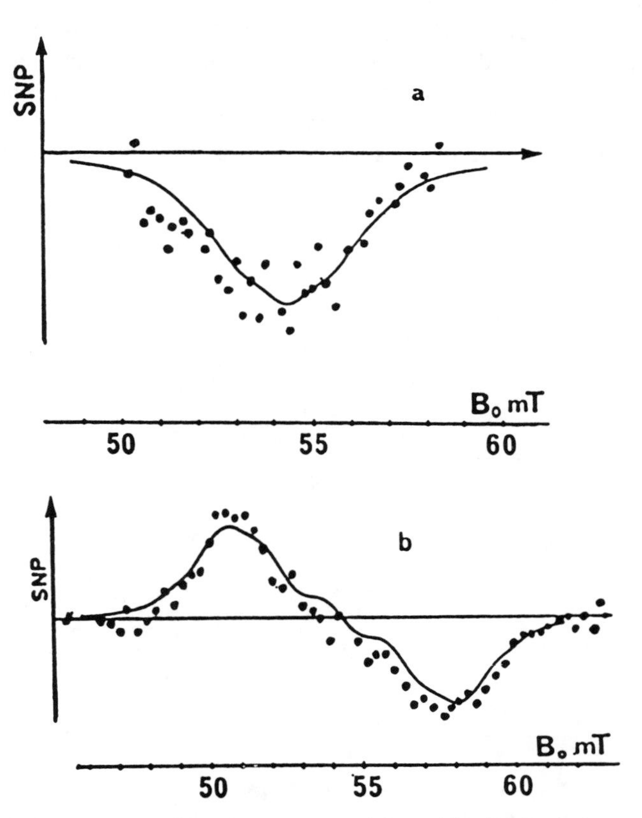

Figure 10.11 SNP spectra of ketones generated from biradicals: *1*, detected by NMR of β-protons (a); *3* detected by ^{13}C NMR of carbonyl carbon atom (b). The frequency of pumping $f = 1531$ MHz, $B_1 = 0.8$ mT (a), 0.45 mT (b) [10].

10.3. SNP in Micellized Radical Pairs

Being a beautiful physical phenomenon, SNP also provides a powerful method to investigate chemical reaction mechanisms and detect their radical stages. For illustration we will now discuss the mechanism of deoxybenzoin photolysis in SDS micelles.

The quantum yield of the $PhCOCH_2Ph$ photolysis is very low, the products are formed in almost undetectable quantities. In ^{13}C NMR spectra under photolysis there are no lines except the one belonging to the carbonyl ^{13}C nuclei of the starting ketone. It exhibits strong emission and indicates intensive reversible photochemical processes, which are accompanied by almost complete regeneration of the starting ketone, so the reaction scheme looks like

$$PhCOCH_2Ph \rightleftharpoons X$$

where X is an unknown intermediate responsible for the ketone CIDNP.

Tarasov et al. [11] investigated ^{13}C SNP in this system and observed the intensive SNP spectra (Fig. 10.12) as a superposition of two doublets. The first one (A/E, with splitting 120 G) was shown to belong to the acyl radical $Ph^{13}CO$; the second, poorly resolved, A/E doublet with smaller splitting 30 G was attributed to the radical $Ph^{13}C(OH)CH_2Ph$, the primary product of the excited triplet ketone photoreduction by the SDS molecule (RH). These SNP spectra support the idea of two competing ways of achieving deoxibenzoin photochemical transformation, dissociation, and photoreduction, through two different intermediates X, the radical pairs $(Ph\dot{C}O\ \dot{C}H_2Ph)^T$ and $[Ph\dot{C}(OH)CH_2Ph\ \dot{R}]^T$, respectively. What is more important, both these pairs

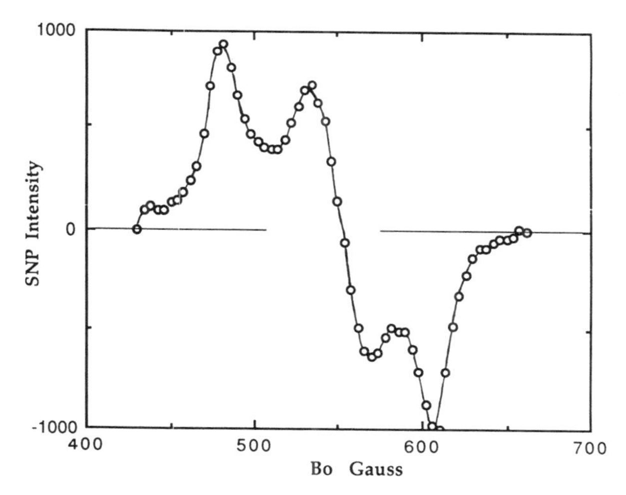

Figure 10.12 Time-resolved SNP spectra taken from the photolysis of deoxybenzoin in SDS micelles.

are highly reactive in ketone regeneration reactions although their kinetic behavior pictured in Fig. 10.13 is different. The recombination of the former pair is much faster than the disproportionation of the latter [11].

Micelles as the microreactors with spatially restricted diffusion of RP partners create favorable conditions for the mw pumping and, hence, for SNP spectroscopy. The RP lifetime in micelles (10–100 ns) is comparable with the intrinsic time of mw-induced spin transition $(\gamma B_1)^{-1} \approx 10$–100 ns. It provides the pair with ample time to respond to mw pumping and results in high SNP intensity. This is clearly seen from the comparison of the two SNP spectra in Fig. 10.3: in micelles the intensity of the SNP spectrum even at ^{13}C natural abundance in ketone is much higher than that in benzene with ketone 45% enriched by ^{13}C (the qualitative measure of the large difference in SNP intensities is that in scattering the points for these two spectra).

The micelle as a microreactor with reflecting surface boundary locks the RP in restricted space. Therefore, the interradical exchange interaction energy J in micellized RP should be a function of the micelle sized, so that the SNP spectra in micelles are the probes of J in RP.

A beautiful illustration of this statement was given by Tarasov and Bagryanskaya [12] for the photolysis of α-methyldeoxybenzoin (99% abundance of ^{13}C in carbonyl site) in micelles of different size, the latter being regulated by the number of carbon atoms n in the detergent molecule $CH_3(CH_2)_{n-1}SO_4Na$. The SNP spectra shown in Fig. 10.14 were found to strongly change as n diminishes. The line broadening and decrease in doublet splitting are the symptoms of J enlargement. Only in SDS micelles ($n = 12$) is J negligibly small

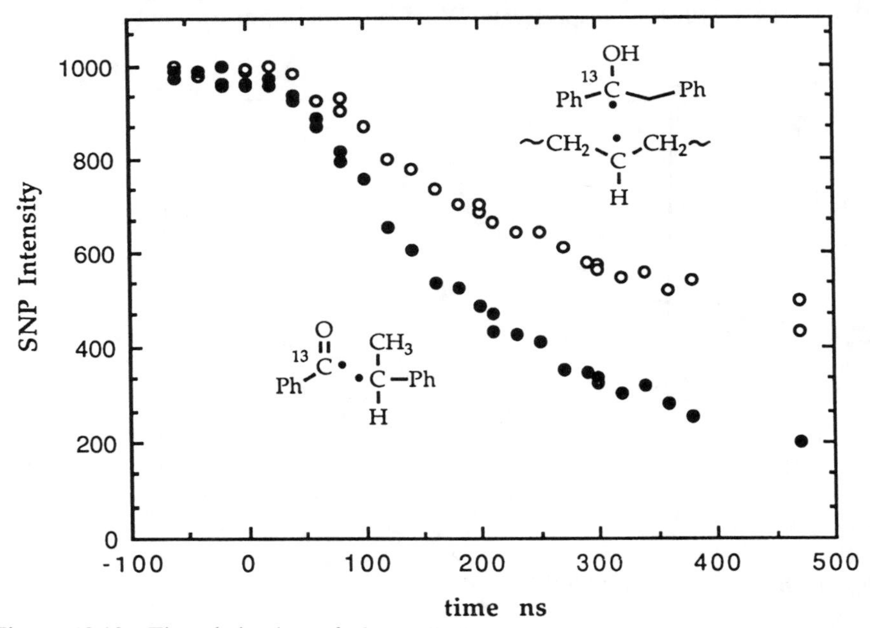

Figure 10.13 Time behavior of the radical pairs generated by photodissociation (bottom) and photoreduction (top) of deoxybenzoin in SDS micelles.

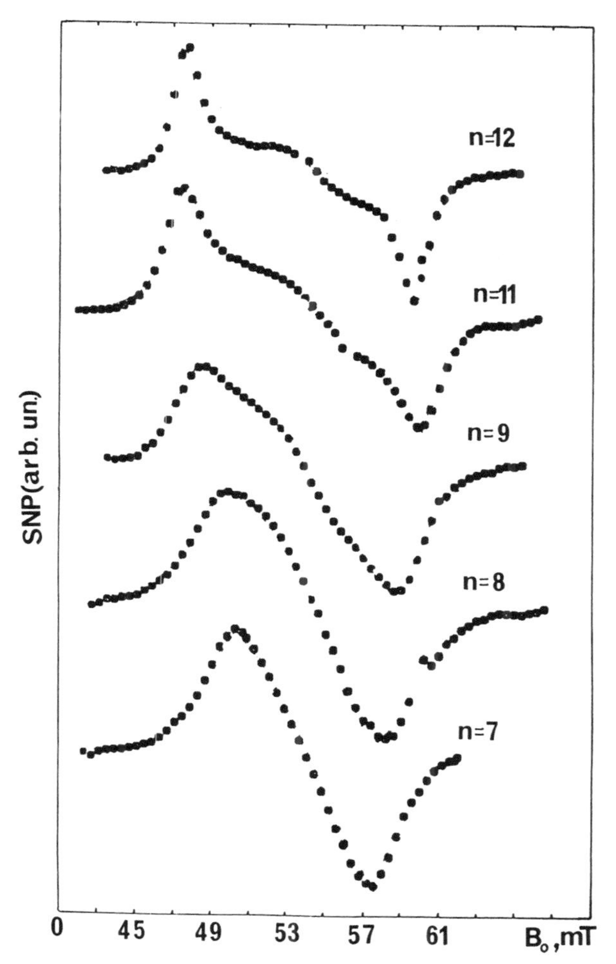

Figure 10.14 SNP spectra from the photolysis of methyldeoxybenzoin in alkyl sulfate micelles detected by ^{13}C NMR of the carbonyl atom; n is the number of carbon atoms in detergent molecule, $f = 1530$ MHz, $B = 0.1$ mT.

and the SNP line splitting corresponds to the HFI constant (124 mT) in the PhCO radical. For smaller micelles and hence for shorter interradical distances J becomes noticeably larger. Below the SNP doublet splittings a and J are given as a function of n:

n	a(mT)	J(mT)
12	12.4	<0.1
11	12.0	0.5
9	10.0	3.4
8	8.4	8.0
7	7.6	14.0

Such information seems to be important for reconstructing the potential energy surfaces for interradical coupling and for calculating the spin dynamics and the magnetic effects more exactly.

The strategy of magnetic probing of exchange potential (SNP, in fact, is one means of probing) along the reaction coordinate of formation or dissociation of the chemical bond is an important supplement to the transition state spectroscopy based on femtosecond time-resolved optical spectroscopy developed recently by Zewail [13]. The difference is that femtosecond spectroscopy probes the transition state of the dissociation reaction at exchange potentials $J \gg kT$, while magnetic probing deals with the exchange potential at long interradical distances and, therefore, at $J \ll kT$. The latter can be considered a transition state radiospectroscopy in analogy with Zewail's optical transition state concept [14].

10.4. Time-Resolved SNP

In time-resolved (TR) SNP experiments the laser and mw pulses are separated by the time delay τ_d. In contrast to the stationary SNP detection, which implies that the RP is affected by the mw field B_1 for their entire lifetime, TR SNP offers a new possibility of mw field influence on the ensembles of RP with lifetimes in selected time domain.

The time resolution (i.e., the shortest τ_d achieved in the TR SNP experiments) was mentioned at the beginning of this chapter to be about 30 ns. Therefore, TR SNP spectroscopy is able to monitor and discriminate the time behavior of the RP ensembles with lifetimes $\geqslant$30 ns. The important advantage of the TR SNP is the ability to separate in time scale the geminate and diffusion radical pairs, again by monitoring SNP spectra as the function of time delay τ_d.

An excellent illustration of TR SNP chemical applications is the photolysis of benzoyl peroxide in CD_3OD [15]. This reaction is known to include the sequence of singlet RPs; one of them ($Ph\dot{C}O_2$ $\dot{P}h$) is specifically interesting because it keeps an active phenyl radical that cannot be observed by conventional ESR technique under standard conditions. The SNP spectra taken with $\tau_d = 0$ (Fig. 10.15a) demonstrate the opposite signs of SNP for the cage (phenylbenzoate) and escape (benzene) products in excellent agreement with the predictions of the reaction scheme and SNP mechanism. However, the TR SNP spectra taken with time delay $\tau_d > 50$ ns were revealed to be completely inverted for both products. In the time domain 50–200 ns, wherein the inverted TR SNP spectra were observed, the SNP is generated in the diffusion RPs such as ($\dot{P}h$ $\dot{P}h$) or ($\dot{P}h$ $\dot{C}D_2OD$), while at times <50 ns the SNP is completely attributed to the geminate RP ($Ph\dot{C}O$ $\dot{P}h$) in the singlet state. The phase inversion dependence on τ_d is depicted in Fig. 10.15b and clearly shows two time domains corresponding to SNP in geminate and diffusion RP ensembles.

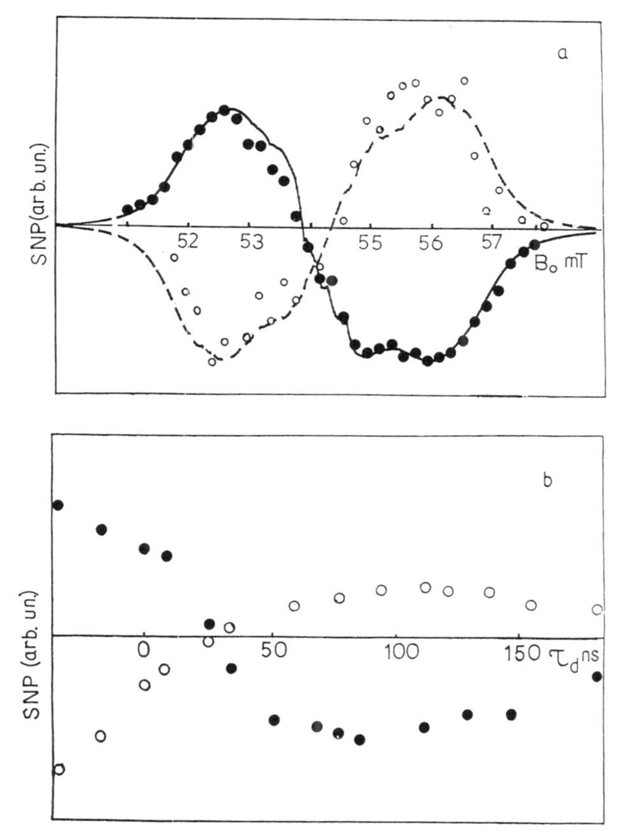

Figure 10.15 Time resolved SNP spectra of phenylbenzoate (filled circles) and benzene (open circles) detected by NMR of ^{13}C nuclei in the photolysis of benzoyl peroxide; $B_1 = 0.6$ mT, $f = 1530$ MHz, $\tau_d = 0$ (a) and their phase inversion as a function of τ_d for phenylbenzoate (filled circles) and benzene (open circles) (b).

To summarize, we conclude that TR SNP spectroscopy is an elegant and powerful technique of chemical physics allowing a real-time probe of radical pair chemical evolution.

References

1. Sagdeev, R. Z., Bagryanskaya, E. G. *Pure Appl. Chem.* **1990**, *62*, 1547.
2. Shkrob, I. A., Tarasov, V. F., Bagryanskaya, E. G. *Chem. Phys.* **1991**, *153*, 427.
3. Bagryanskaya, E. G., Grishin, Yu. A., Sagdeev, R. Z., Leshina, T. V., Polyakov, N. I. *Chem. Phys. Lett.* **1985**, *117*, 220.
4. Meng, K. Q.-X., Suzuki, K., Terazima, M., Azumi, T. *Chem. Phys. Lett.* **1990**, *175*, 364.
5. Kruppa, A. I., Leshina, T. V., Sagdeev, R. Z. *Chem. Phys. Lett.* **1985**, *121*, 386.

6. Tsentalovich, Yu. P., Bagryanskaya, E. G., Grishin, Yu. A., Sagdeev, R. Z., Roth, K. H. *Chem. Phys.* **1990**, *142*, 75.

7. Bagryanskaya, E. G., Grishin, Yu. A., Sagdeev, R. Z. *Chem. Phys. Lett.* **1986**, *128*, 417.

8. Michailov, S. A., Salikhov, K. M., Plato, A. M. *Chem. Phys.* **1987**, *117,* 197.

9. Kaptein, R. *J. Am. Chem. Soc.* **1972**, *94*, 6251.

10. Koptyug, I. V., Bagryanskaya, E. G., Grishin, Yu. A., Sagdeev, R. Z. *Chem. Phys.* **1990**, *145*, 375.

11. Tarasov, V. F., Bagryanskaya, E. G., Turro, N. J. *Chem. Phys.* **1993**, *209*, 409.

12. Tarasov, V. F., Bagryanskaya, E. G. *Chem. Phys.* **1993**, *174*, 237.

13. Zewail, A. H. *Science* **1988**, *242*, 1645.

14. Buchachenko, A. L., Tarasov, V. F., Ghatlia, N. G., Turro, N. J. *Chem. Phys. Lett.* **1992**, *192*, 139.

15. Avdievich, N. A., Bagryanskaya, E. G., Grishin. Yu. A., Sagdeev, R. Z. *Chem. Phys. Lett.* **1989**, *155*, 141.

CHAPTER

11

Coherence in Spin Dynamics and Chemical Reactivity

Precessive electron spin motion results in a time oscillating population of singlet and triplet states of radical pair and therefore in oscillation of its chemical reactivity. Providing that the radical pairs are prepared instantly and start their spin evolution simultaneously the oscillation in chemical reactivity is followed by oscillation of the yield of chemical reaction products detected by measuring fluorescence intensity, conductivity, etc. The decay of simultaneously prepared RPs in the spin state of definite multiplicity is modulated by these oscillations and results in the periodicity, and the quantum beats in the yield of RP reaction products.

The coherence of spin dynamics results in the coherence of chemical reactivity. It is the first and to date unique situation when coherence, as a physical property of molecular systems, manifests itself in chemical reactions.

RP spin dynamics were shown in Chapter 2 to predict oscillations in singlet state population according to $\sin^2(Q_{ab}t)$, where Q_{ab} is a frequency of triplet–singlet transitions, equal to the difference of the Larmor frequencies of the two unpaired electron spins of the RP partners. To inspect quantum beats it is necessary to solve the kinetic equations for the time-dependent spin state populations taking into account the consecutive reencounters of the partners, the exchange interaction in the partner contacts, the lifetime of the RP, the duration of contacts, the time interval between contacts, and the recombination rate constants. For high magnetic fields this problem has been solved quantitatively by Luders and Salikhov [1] for the particular case of one-frequency RP (i.e., the pair with single Larmor frequency difference Q_{ab}). The calculated singlet state population ${}^{T}\rho_{SS}$ for triplet RP as a precursor is shown in Fig. 11.1.

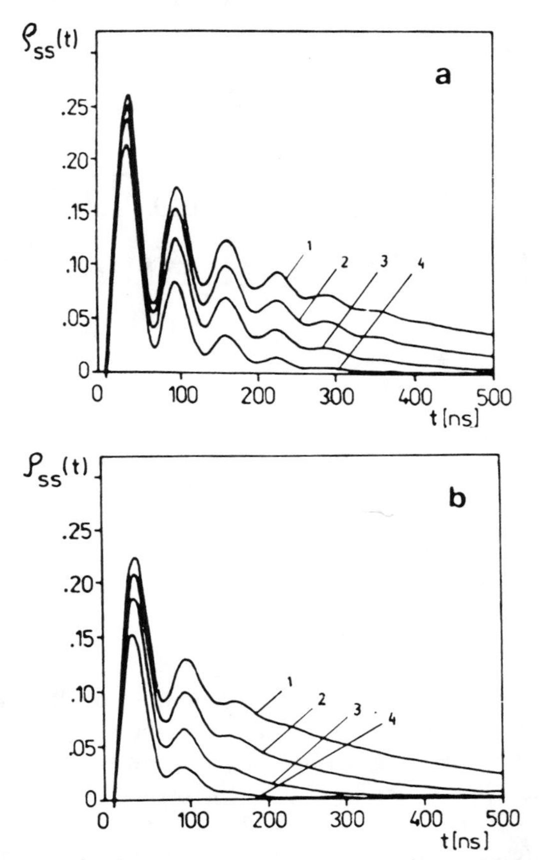

Figure 11.1 Time oscillations of the singlet state population for the triplet precursor. Parameters: $\omega_{ST_0} = 10^{-8}\,s^{-1}$, $\tau = 500$ ns, $p_S = 0.1$ (1), 0.25 (2), 0.5 (3), 1.0 (4); $\tau_2 = 50$ ns (a), 25 ns (b). ω_{ST_0} is the frequency of triplet–singlet transitions equal to the difference of Larmor frequencies of electron spins of RP partners, τ is the RP lifetime, p_S is the probability of radical recombination in the singlet contact pair, τ_2 is the mean time interval between two consecutive contacts of radicals in the pair. The damping does not (or only slightly) depend on the p_S. It means that those contacts during which the partners do not recombine produce the same phase mixing as those that are accompanied by recombination.

The clearly seen damping oscillations in the population of reactive singlet state are more pronounced the longer the time period between consecutive contacts, the time that the pair used for triplet–singlet evolution.

Now after such a preliminary inspection we are ready to analyze the quantum beats in chemical systems.

There are two types of quantum beats: the native ones induced by internal magnetic interactions (hyperfine or Zeeman) and those stimulated by external microwave magnetic fields.

11.1. Native Quantum Beats

Klein and Voltz [2, 3] were the first to discover chemical reactivity coherence. They observed the quantum beats in the decay of radioluminescence induced by fast electron pulse irradiation of solutions of the scintillator molecules (M) in cyclohexane (S). The sequence of reactions

$$\mathrm{S} \rightarrow \dot{\mathrm{S}}^{+} + e$$

$$\mathrm{M} + e \rightarrow \dot{\mathrm{M}}^{-}$$

$$\mathrm{M} + \dot{\mathrm{S}}^{+} \rightarrow \dot{\mathrm{M}}^{+} + \mathrm{S}$$

generates ion radical pair $\dot{\mathrm{M}}^{+}$ and $\dot{\mathrm{M}}^{-}$ locked in an Onsager sphere in the singlet spin state. Spin-selective coupling of ions in the pair

$$(\dot{\mathrm{M}}^{+}\ \dot{\mathrm{M}}^{-})^{\mathrm{S}} \rightarrow \mathrm{M}^{*} + \mathrm{M}$$

generates excited scintillator molecules M* and their fluorescence has been detected by the single photon counting technique.

The time evolution of the relative increase in radioluminescence intensity after the electron pulse, shown in Fig. 11.2, exhibits quantum beats as damped

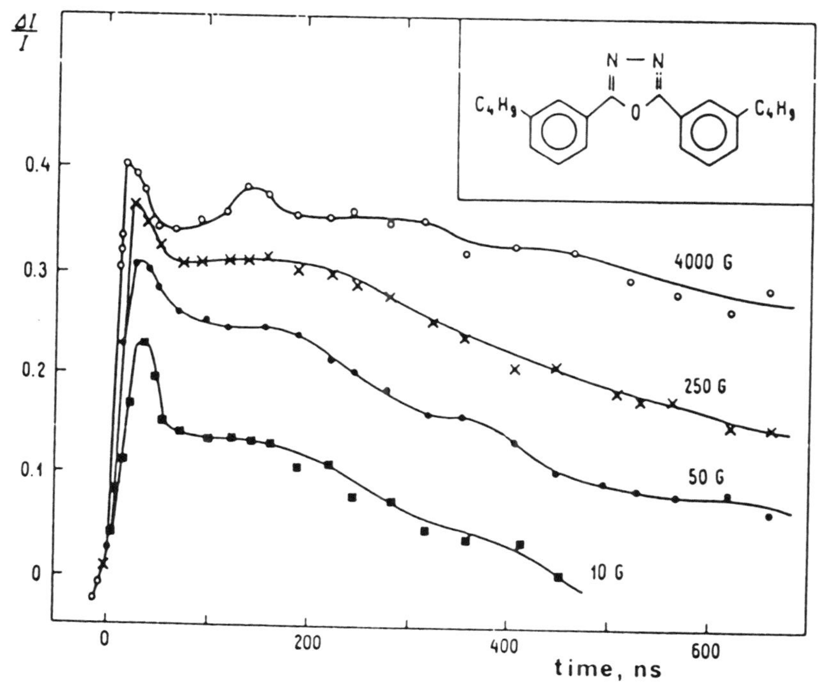

Figure 11.2 Time evolution of the radioluminescence intensity at different magnetic fields in 1.5×10^{-3} mol scintillator solution (see the top formula). The changes of intensity are normalized to the initial intensity I [2].

oscillations. The frequency of these oscillations is about 8 MHz and coincides with the HFI coupling constant in the scintillator's ion radicals and, therefore, with the frequency of singlet–triplet conversion driven by HFI.

Figure 11.2 also demonstrates the growth in luminescence intensity with magnetic field strength as should be expected for the singlet state of the starting ion radical pair. The singlet–triplet conversion in this RP is partly hampered by the magnetic field. As a result, the singlet spin state is partly locked and the RP preferably reacts generating the molecule M*, which emits the fluorescence.

The exponential decay at long times is attributed to the irreversible and nonoscillating processes of spin–spin and spin–lattice relaxation, which are responsible for the leakage of spin evolution of RP through other magnetic interactions competing with HFI and Zeeman coupling.

It was really an advantage for Klein and Voltz [2, 3] to observe the quantum beats in a multinuclear spin system with many superimposed beats frequencies. Perhaps in ion radicals of scintillators studied the HFI coupling constants are in simple ratios and this favors the detection of discrete oscillation frequencies. Klein and Voltz failed to detect distinct beats in solutions of multinuclear molecules such as anthracene and perylene, in accordance with this conclusion.

Direct evidence of chemical reactivity coherence governed by hyperfine and Zeeman interactions was given by Anisimov and colleagues [4, 5]. They observed quantum beats in fluorescence generated by ion radical coupling in solutions irradiated by fast electron pulses. The scheme of chemical reactions is similar to that cited above; the only difference is that the partners of the ion radical pair were not chemically identical.

Well-defined beats are shown in Fig. 11.3. These fluorescence oscillations reflect the singlet–triplet oscillations in the pair of $(\text{tetramethylethylene})^+/(\text{paraterphenyl-d}_{14})^-$ with the starting singlet spin state. The tetramethylethylene radical–cation possesses 12 magnetically equivalent hydrogen nuclei with HFI coupling constant 16.5 G, while in deuterated paraterphenyl radical–anion HFI coupling constants can be ignored, so that the spin evolution of the pair is driven by the single HFI coupling constant and is described by a single frequency about 50 MHz.

Similar oscillations for the same chemical system in zero magnetic field are shown in Fig. 11.4. The parameters of experimental and calculated beats are summarized in Table 11.1.

The coincidence of experimental and theoretical beat periods unambiguously proves that the singlet–triplet coherence induced by HFI is responsible for the quantum beats. However, the experimental beat amplitudes are lower than those calculated theoretically. The discrepancy might be the result, first, of the leakage of the RP spin evolution through other magnetic interactions and, second, of the recombination fluorescence from some other pairs, such as $(\text{paraterphenyl-d}_{14})^+/(\text{paraterphenyl-d}_{14})^-$ and $(\text{solvent})^+/(\text{paraterphenyl-d}_{14})^-$, which smears the oscillations.

For the ion radical pair $(\text{diphenylsulfide})^+/(\text{paraterphenyl-d}_{14})^-$ formed by

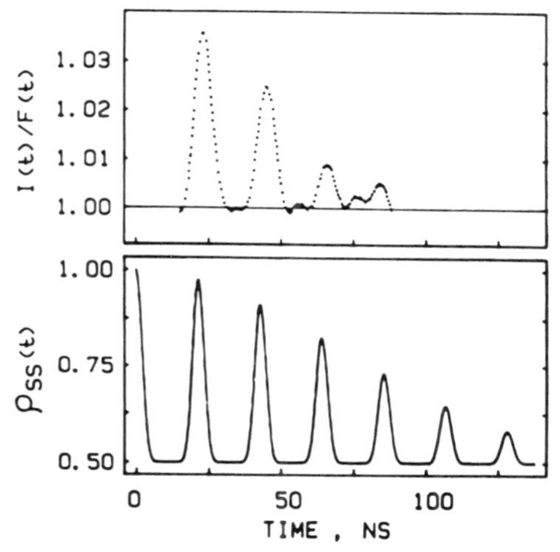

Figure 11.3 Fluorescence intensity oscillations determined as the ratio $I(t)$ of the experimental intensity to the smoothed kinetic function of RP decay that imitates the pair lifetime distribution. Top: experimental beats for (tetramethylethylene)$^+$/(paraterphenyl-d_{14})$^-$ in *trans*-decalin. Bottom: the calculated singlet spin state populations in high magnetic field [4].

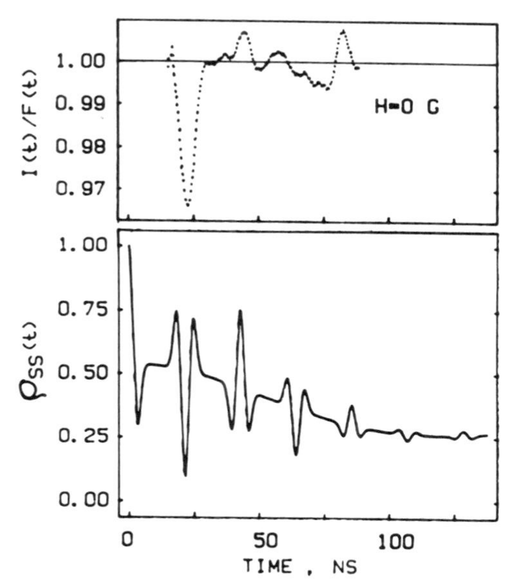

Figure 11.4 Fluorescence intensity oscillations similar to those in Fig. 11.3 for zero magnetic field [4].

Table 11.1 Experimental and Calculated Parameters for the Quantum Beats in the Recombination of Radical Ion Pairs ($H = 3300$ G)

Radical ion pair	Solvent	HFI constant a(G)	Beat period (ns) Calc. $\tau = 2\pi/\gamma a$	Exp.
(Tetramethylethylene)$^+$/ (paraterphenyl-d_{14})$^-$	*trans*-Decaline	16.5	21.6	22 ± 0.5
	cis-Decaline	—	—	22 ± 0.5
	Squalane	16.6	21.5	22 ± 0.5
	n-Pentadecane	—	—	22 ± 0.5
(Durene)$^+$/parater-phenyl-d_{14})$^-$	*trans*-Decaline	10.8	33.1	33.5 ± 0.5

radiolysis of diphenylsulfide and paraterphenyl-d_{14} solutions in *cis*-decaline the leading role in singlet–triplet evolution belongs to the difference in Zeeman energy of partners $\Delta g\beta H$, since the HFI coupling constants are rather small in both partners. The formation of the paraterphenyl-d_{14} molecules in the excited singlet state as the reaction product of the singlet spin state ion radical pairs was inspected by detecting their fluorescence, $\tau_{fl} \approx 1$ ns [5].

The fluorescence intensity oscillations are clearly seen even on the decay curve (Fig. 11.5). The difference between the observed fluorescence intensity decay curve and the smoothed one provides the picture of pure quantum beats, shown in Fig. 11.6, together with the singlet spin state population $\rho_{SS}(t)$ calculated with the $\Delta g\beta H$ term only.

The in-phase behavior of the experimental and theoretical beats proves certainly that the spin evolution of the ion radical pair is directed exclusively by the Zeeman term $\Delta g\beta H$. This conclusion is illustrated in Fig. 11.7, where the beats frequencies $1/T$ (T is the beats period) are plotted as a function of the magnetic field. The linearity of this dependence confirms that the Zeeman interaction is responsible for the coherence of the ion radical pair chemical reactivity.

The quantum beats can be considered as an elegant method for investigating the mechanisms of chemical reactions including those of biophysical importance.

The photoexcitation of the donor P_{700} in plant photosystem I induces the following electron transfer sequence:

$$P^*_{700}A_0A_1(FeS) \rightarrow P^+_{700}A_0^-A_1(FeS) \rightarrow P^+_{700}A_0A_1^-(FeS) \rightarrow P^+_{700}A_0A_1(FeS)^-$$

where A_1 is the vitamin K_1 molecule. The intermediate radical pair $P^+_{700}A_0^-A_1(FeS)$ generated by laser pulse in fully deuterated whole cells of cyanobacterium *Synechococcus lividus* was found to give transient ESR spectra with polarized unpaired electron spins [6]. The time evolution of the transverse electron spin magnetization was studied for various static and microwave

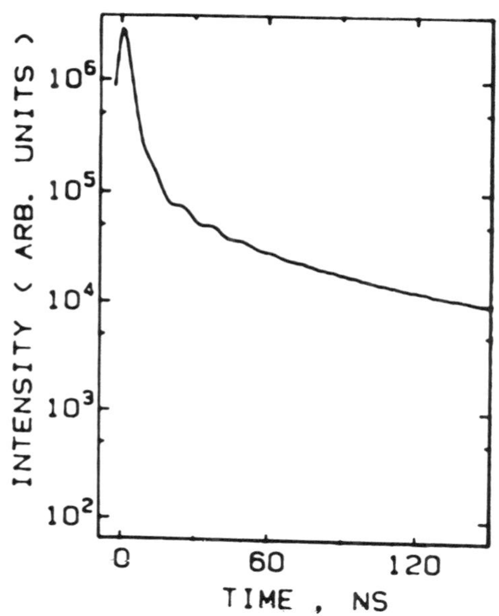

Figure 11.5 The fluorescence decay curve after electron pulse irradiation of mixed solution of 6×10^{-2} mol diphenylsulfide and 10^{-3} paraterphenyl-d_{14} in *cis*-decaline. $H = 12\,\mathrm{kG}$ [5].

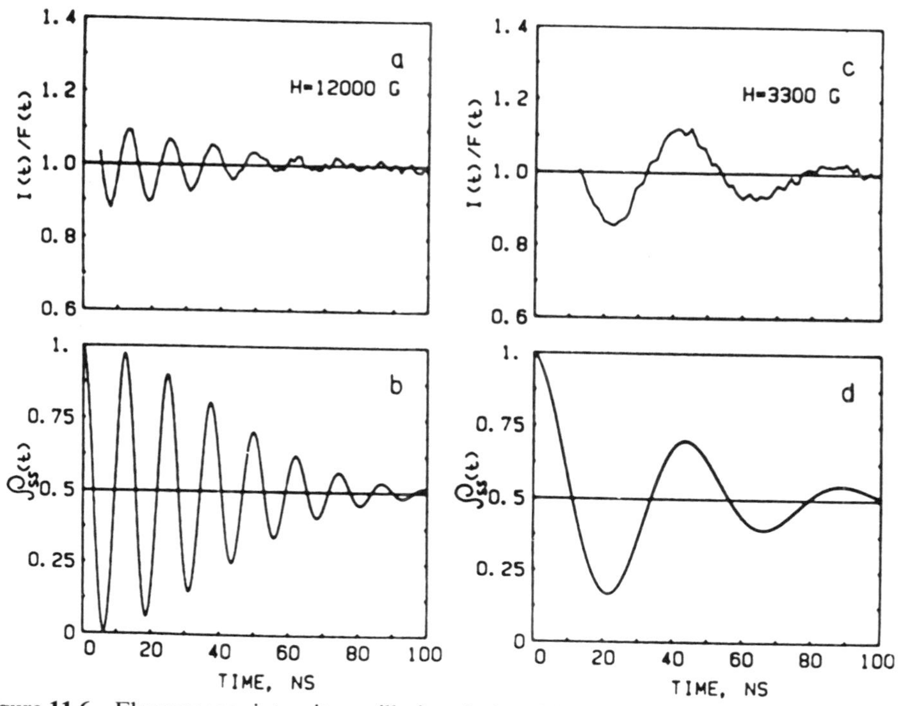

Figure 11.6 Fluorescence intensity oscillations induced by recombination of (diphenylsulfide)$^+$/(paraterphenyl-d_{14})$^-$ pairs in *cis*-decaline. (a, c) Experiment, (b, d) theory; (a, b) $H = 12\,\mathrm{kG}$, (c, d) $H = 3\,\mathrm{kG}$.

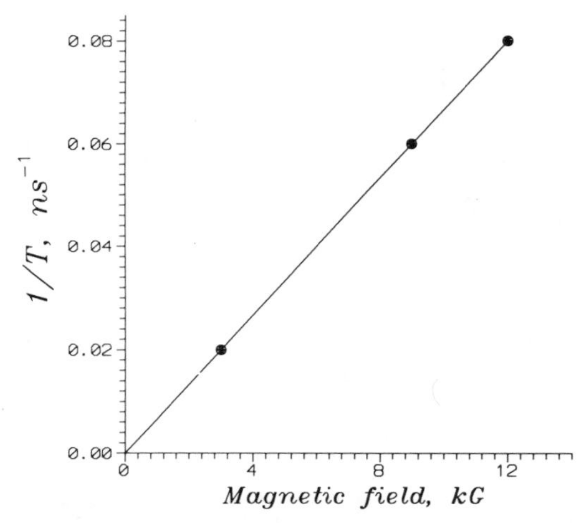

Figure 11.7 The linear relation between the quantum beats frequency and magnetic field strength [5].

magnetic fields. At times early after laser excitation the time profiles of the spectra were found to reveal the strongly expressed oscillatory behavior damping at long time. The oscillations represent quantum beats associated with the spin-correlated generation of the radical pairs in plant photosystem I and coherent behavior of the electron spin magnetization.

The detection of quantum beats in photosynthetic reaction centers provides direct confirmation of the coherent spin states in electron transfer photoreactions. Furthermore, the frequencies of the beats and their amplitude variations shown in Fig. 11.8 as a two-dimensional pattern [7] provide direct measurement of a number of important magnetic and kinetic parameters of the centers. The calculated quantum beats were found to nicely reproduce the experimental ones with magnetic parameters g_x, g_y, g_z being equal to 2.00285, 2.00285, and 2.0022 for P_{700}^{+} and 2.0055, 2.0046, and 2.0023 for A_1^-, respectively, in accordance with those determined by stationary ESR spectroscopy. Even mutual orientations of the partners as well as their spin–spin coupling parameters $J = 0$, $D = -0.12\,\text{mT}$, $E = 0$ were confirmed by quantum beats studies. Chemical decay and relaxation parameters were measured from oscillation damping to be 250 ns, the lifetime of the radical pair, $T_1 > 1\,\mu\text{s}$, $T_2 = 500\,\text{ns}$.

The important conclusion derived from quantum beats studies was that in native photosystem I the secondary RP $P_{700}^{+}A_1^-$ is generated in a virtually pure singlet state. All these results provide evidence that quantum beats are sensitive dynamic probes for the primary reactions in photosynthesis as well as in photochemistry.

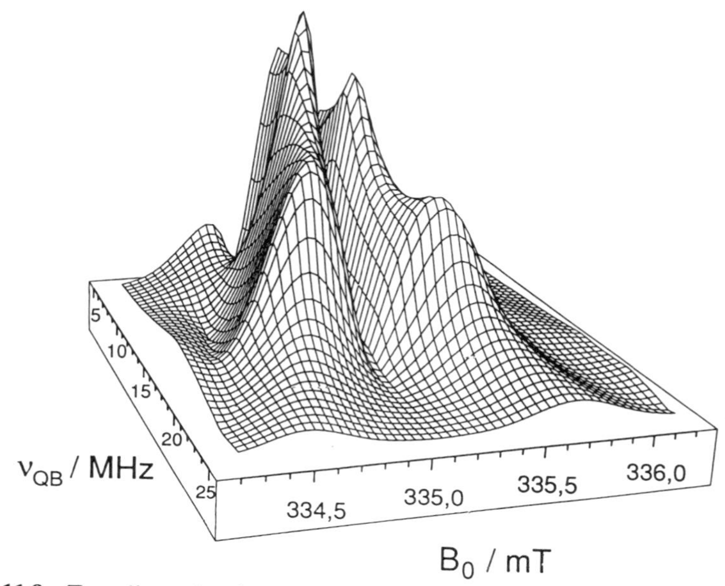

Figure 11.8 Two-dimensional presentation of quantum beats amplitudes at the different beats frequencies ν_{QB} as a function of magnetic field B_0 [7].

11.2. Microwave-Stimulated Quantum Beats

The quantum beats described in the previous section are inherent to the chemically reacting system, radical or ion radical pair, because they are induced by the difference in local, internal magnetic fields experienced by each partner of the pair. In general, there is a collection of superimposed beats with individual frequencies

$$\Delta\omega_{ij} = \frac{1}{2}\left[(g_a - g_b)\beta H + \sum_i a_i m_i^a - \sum_j a_j m_j^b\right]$$

where g_a and g_b are g-factors of the RP partners a and b, a_i, a_j, m_i^a, and m_j^b are the HFI constants and nuclear spin projections for i and j sorts of nuclei (see Chapter 2), and $\Delta\omega_{ij}$ is the frequency of the singlet–triplet oscillations for the ij-ensemble of radical pairs.

Microwave fields resonant with external magnetic field were predicted theoretically [8] to promote periodic variations in the population of the RP singlet state and therefore to stimulate quantum beats in chemical reactivity.

Figure 11.9 illustrates the origin of stimulated beats in terms of a semiclassical vector model. The spins of the RP partners are shown by vectors $\mathbf{S}_1$ and $\mathbf{S}_2$. Let us use a coordinate system rotating with frequency ω, the precession frequency of spins $\mathbf{S}_1$. In the rotating frame spin $\mathbf{S}_1$ is motionless, while the spin

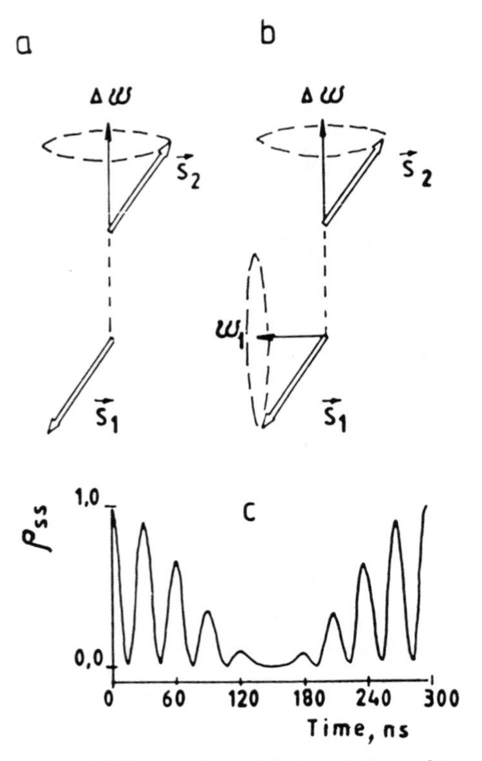

Figure 11.9 The vector model illustration of the evolution of two spins $\mathbf{S}_1$ and $\mathbf{S}_2$ in rotating coordinate system in absence of mw field H_1 (a), with a small mw field amplitude (b), and spin population of the RP singlet state as a function of time (c). $\Delta\omega$ is the native beats frequency ($\Delta H = \Delta\omega/\gamma = 12\,\mathrm{G}$), ω_1 is a frequency of microwave-induced beats ($H_1 = \omega_1/\gamma = 1.2\,\mathrm{G}$) [9].

$\mathbf{S}_2$ precesses with detuning frequency $\Delta\omega$ resulting in a population oscillation between the T_0 and S states, or to the native quantum beats.

The microwave (mw) field H_1 with frequency ω_1 resonant with spin $\mathbf{S}_1$ causes its precession and flips between the T_0 and T_+ states with the frequency $\omega_1 = \gamma H_1$. For the case $\omega_1 \ll \Delta\omega$ the fast S–T_0 oscillations equilibrate the populations of the S and T_0 states, while slow flips between the T_0 and T_+ states result in synchronized depopulation of the S and T_0 states. In other words, the yield of chemical products is doubly modulated, first, by native, high-frequency oscillations with the period $\Delta\omega^{-1}$ and, second, by mw field-induced slow oscillations, or mw-induced quantum beats with a period $(\gamma H_1)^{-1}$.

Figure 11.9 shows the example of such two-frequency beats, the first with $\Delta\omega^{-1} = 30\,\mathrm{ns}$ and the second with $\omega^{-1} = (\gamma H_1)^{-1} = 300\,\mathrm{ns}$ induced by mw field $H_1 = 1.2\,\mathrm{G}$.

Experimentally the stimulated quantum beats were first detected by Saik et al. [9] at the radiolysis of paraterphenyl-h_{14} solutions by fast electrons. Figure 11.10 demonstrates the data obtained by subtracting two kinetic curves: the first corresponds to recombination luminescence with mw field $H_1 = 8.0\,G$ in the resonant external magnetic field $H = 34\,G$, and the second is a similar luminescence curve in the absence of an mw field. The signal difference is normalized to the amplitudes of the second curve (top part of Fig. 11.10).

The beats period is seen to be 46 ± 5 ns in quantitative agreement with the magnitude of $(\gamma H_1)^{-1} = 44$ ns. Other evidence is that the beats were absent if the mw field H_1 was not resonant (external magnetic field $H = 50\,G$ instead of 34 G as for resonance conditions).

At last, the beats period T was shown to follow exactly the amplitude of mw field H_1 (Fig. 11.11). According to the ratio $T^{-1} = \omega_1 = \gamma H_1$ the slope of the linear dependence on Fig. 11.11 should be equal to γ, $2.8 \times 10^6\,s^{-1}\,G^{-1}$, in excellent agreement with the experimental value, $(2.85 \pm 0.1) \times 10^6\,s^{-1}\,G^{-1}$.

Microwave-stimulated quantum beats were observed also in the 10^{-3} mol solution of paraterphenyl-d_{14} in squalane. Due to smaller HFI constants in comparison with those of paraterphenyl-h_{14} the beats were detected only for $H_1 \leqslant 4\,G$ to fulfill the ratio $\omega_1 \ll \Delta\omega$ [9].

Strong beats were easily detected in 10^{-2} mol solutions of paraterphenyl-d_{14} and tetramethylethylene in squalane up to $H_1 = 10\,G$; however, for biphenyl 10^{-1} mol solutions in squalane the beats have not been observed. This was

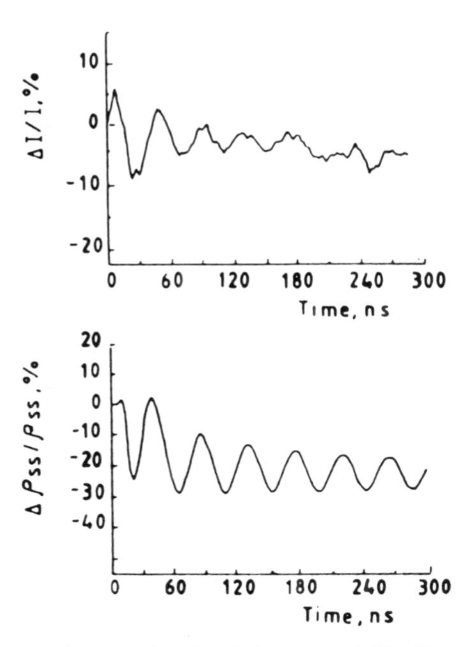

Figure 11.10 Quantum beats stimulated by mw field. Top: experiment for the 10^{-3} mol paraterphenyl-h_{14} solution in squalane, $H_1 = 8\,G$. Bottom: the calculated population of singlet spin state for pair $(\text{paraterphenyl-}h_{14})^+/(\text{paraterphenyl-}h_{14})^-$ [9].

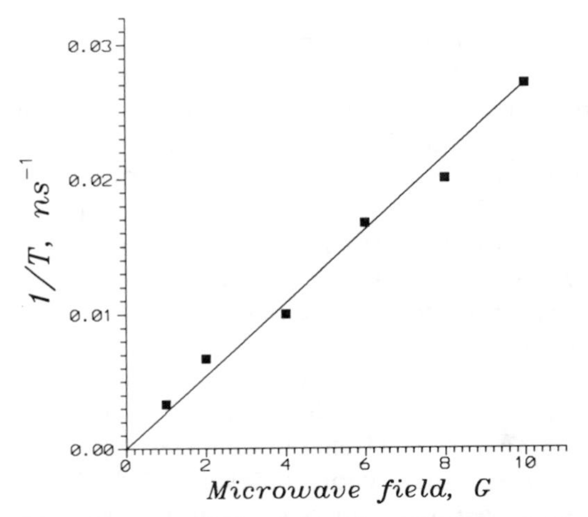

Figure 11.11 The beats frequency $1/T$ as a function of the mw field amplitude H_1 [9].

attributed to the fact that at high concentrations biphenyl undergoes fast ion-molecular charge transfer that narrows the ESR line and shortens dephasing relaxation time T_2 to the magnitude <10 ns, which is too short to observe stimulated presession coherency at the experimentally available mw fields H_1.

References

1. Luders, K., Salikhov, K. *Chem. Phys.* **1989**, *134*, 31.
2. Klein, J., Voltz, R. *Phys. Rev. Lett.* **1976**, *36*, 1214.
3. Klein, J., Voltz, R. *Can. J. Chem.* **1977**, *55*, 2103.
4. Anisimov, O. A., Bizyaev, V. L., Lukzen, N. N., Grigoryants, V. M., Molin, Yu. N. *Chem. Phys. Lett.* **1983**, *101* 131.
5. Veselov, A. V., Melekhov, V. I., Anisimov, O. A., Molin, Yu. N. *Chem. Phys. Lett.* **1987**, *136*, 263.
6. Kothe, G., Weber, S., Bittl, R., Ohmes, E., Thurnauer, M., Norris, J. *Chem. Phys. Lett.* **1991**, *186*, 474.
7. Kothe, G., Weber, S., Bittl, R., Ohmes, E., Thurnauer, M., Norris, J. Quantum beats as probes of the spin dynamics in photosynthesis. In *Magnetic Field Effects and Spin Effects in Chemistry*, book of abstracts. Konstanz, Germany, 1992.
8. Salikhov, K., Molin, Yu., Sagdeev, R., Buchachenko, A. *Spin Polarization and Magnetic Effects in Radical Reactions.* Elsevier, Amsterdam, 1984.
9. Saik, V. O., Anisimov, O. A., Koptyug, A. V., Molin, Yu. N. *Chem. Phys. Lett.* **1990**, *165*, 142.

CHAPTER

12

On the Action of Electromagnetic Waves on Biological Processes

The study of interactions of nonionizing electromagnetic waves with biologically important systems (macromolecules, molecular assemblies, cells, whole organs, and organisms) has attracted much attention as an important part of the general research program that covers the electromagnetic field (EMF)–matter interactions. Additionally, and perhaps the most stimulating factor, is the public interest in the possible adverse health effects from EMF generated, for example, by 50 Hz high-voltage transmission lines, video displays, clinic NMR imaging procedures, etc. Particular concern has come from epidemiological findings that correlate the exposure of humans to weak EMFs with elevated risk for developing health pathologies and medical deviations. The growing number of experimental observations of the biological and medical effects of EMF has been reported during the past decades and hypothetical mechanisms that might be involved in mediating these effects have been suggested (for review, see [1–6]).

The purpose of this chapter is to discuss the physical grounds and origins of the biological effects of EMF rather than to estimate which of them are worthy of confidence and which seem to be unreliable. First, it is necessary to keep in mind that both EMF components—electric and magnetic—might be responsible for the biological effects of EMF. Based to this undisputable statement two concepts should be considered a priori. The first one concerns the electric component that interacts with the dielectric part of the biological systems and induces dielectric losses and the rise of temperature resulting in modification of bioprocesses (enzyme activity, membrane-mediated signal transduction, DNA and RNA synthesis, Ca^{2+} regulation, etc.) and their macroscopic manifestations (the rate of the cell growth and other human biological responses). The alternative concept concerns the magnetic EMF component that is supposed to interact with paramagnetic intermediates (such as radical or ion radical pairs) and result in spin-mediated mechanisms similar to those described in Chapters 4–11. Both concepts offer theoretically feasible pathways by which relatively weak EMFs could affect biological functions.

The latter concept prevailed during the past decade mostly under the powerful influence of impressive ideas and the results of spin chemistry. The great temptation to explain the biological effects of EMF in terms of *electromagnetic spin chemistry* resulted in strong exaggeration of the biological importance of the spin-related mechanisms. Indeed, to be measurably influenced by EMFs the biological systems, both in vivo and in vitro, are required to possess paramagnetic intermediates with the spin evolution in the nanosecond time domain and with spin-selective chemical transformations. The best and most reliable probe for the existence of such intermediates is the observation of first-generation magnetic effects (MFE, MIE, CIDNP, CIDEP). Since the discovery of these effects numerous efforts were made to observe them in various biochemical systems and processes, however, the results were more than modest (the only remarkable exception is photosynthesis). Neither CIDNP nor CIDEP was found in biochemical reactions; MFE of small magnitude was observed only in a very restricted number of biochemical processes (see Chapter 1).

This provides strong evidence that in biochemical processes there are no radical or ion radical pairs with characteristics dynamic and magnetic enough to provide significant magnetic effects and therefore be influenced and chemically modified by EMFs. Even if they exist as intermediates they probably have lifetimes or relaxation times too short (for instance, pairs with Fe^{2+} or Fe^{3+} ions as partners) for the EMF to be able to interfere into their spin dynamics.

These arguments favor the conclusion that the priority in the stimulation of biological effects of EMF belongs to the electric EMF component. Of key importance is a structural microheterogeneity at the level of cell, organ, or organism, which has two important consequences. First, it results in the nonhomogeneous distribution of the EMF in structural elements of biosystems, so that local EMF amplitudes might exceed averaged ones by several orders of magnitude. Second, the nonequivalence of the various structural elements of cells and organs with respect to dielectric and conducting properties provokes strongly nonhomogeneous absorption of EMF, so that locally induced temperatures and ion currents might exceed the averaged ones again. As a result, being microscopically too small to be registered by standard physical methods, this local heating might appear considerable for cell functions.

Both these factors provide the strongly nonhomogeneous absorption of EMF, which stimulates the local deviations and perturbations in the biological activity of the molecular assemblies, cells, ion channels, membranes, and, finally, the whole organism response. These conclusions are in agreement with the majority of experimental findings. In particular, the biological effects of

EMF were found, as a rule, only in microheterogeneous systems but were not detected in homogeneous solutions that chemically model the biological systems [7].

The authors intend neither to discuss the numerous biological effects of EMF summarized in many monographs [1, 3–5, 7], nor to ignore the spin-related mechanisms of the effects of EMF. They would like to emphasize only that the dominant role in EMF-induced biological and medical effects belongs to the "electric" scenario written by the electric EMF component interacting with biomolecular systems.

References

1. Chiabrera, A., Nicolini, C., Schwan, H. P., eds. *Interactions between Electromagnetic Fields and Cells*. Plenum, New York, 1985.
2. Devyatkov, N. D., Sevastyanova, L. A., Vilenskaya, R. L. *Sol. Phys. Usp.* **1974**, *16*, 568.
3. Polk, C., Postow, E., eds. *CRC Handbook of Biological Effects of Electromagnetic Fields*. CRC, Boca Raton, FL, 1986.
4. Blank, M., Findl, E., eds. *Mechanistic Approaches to Interactions of Electromagnetic Fields with Living Systems*. Plenum, New York, 1987.
5. Wilson, B. W., Stevens, R. G., Anderson, L. E., eds. *Extremely Low Frequency Electromagnetic Fields: The Question of Cancer*. Batelle, Columbus, 1991.
6. Grundler, W., Kaiser, F., Keilmann, F., Walleczek, J. *Naturwissenschaften* **1992**, No. 1728, 1.
7. Akoev, I. G., ed. *Biological Effects of Electromagnetic Fields*. Biophysics Press, Puschino, 1986.

Conclusion

Two remarkable radiophysical properties of chemical reactions—to act as an emitter, the chemical raser, and to be a receiver of electromagnetic waves—have been elucidated in this book. They both originate from the electron and nuclear spin selectivity of the chemical interaction in the reagent pairs whose partners are paramagnetic species (radicals, ions, triplet molecules, etc.). Such a pair should be considered as a spin-selective chemical microreactor whose spin and, therefore, chemical behavior are controlled by magnetic interaction, the main component of the chemical reaction magnetic scenario.

Electromagnetic field influences on spin dynamics is a new component of the electromagnetic scenario. Such an intervention results in new remarkable phenomena in the chemical reactions described in this book and summarized schematically below.

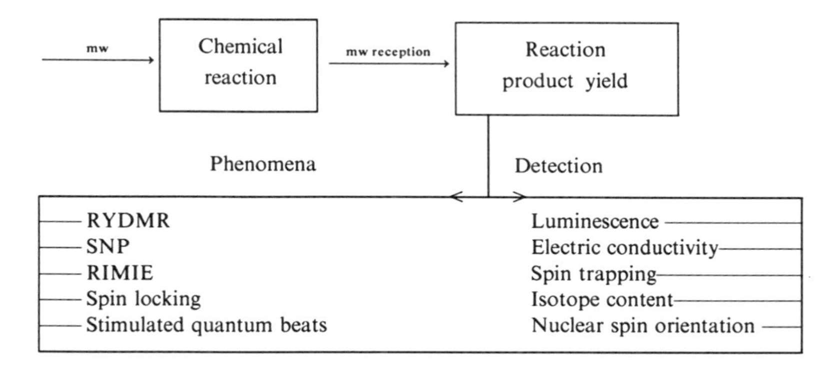

The authors are grateful to the readers of this book and hope that it will encourage them not only to use the achievements of electromagnetic spin chemistry but to further advance it updating the above scheme with new fascinating phenomena.

Index